The Unnatural Trade

The Unnatural Trade

SLAVERY, ABOLITION, AND ENVIRONMENTAL WRITING, 1650–1807

Brycchan Carey

Yale UNIVERSITY PRESS

New Haven & London

Published with assistance from the Annie Burr Lewis Fund.
Published with assistance from the foundation established in memory of Philip Hamilton McMillan of the Class of 1894, Yale College.

Yale University Press books may be purchased in quantity for educational, business, or promotional use. For information, please e-mail sales.press@yale.edu (U.S. office) or sales@yaleup.co.uk (U.K. office).

Set in Adobe Garamond type by IDS Infotech, Ltd.
Printed in the United States of America.

Library of Congress Control Number: 2024930168
ISBN 978-0-300-22441-2 (hardcover : alk. paper)

A catalogue record for this book is available from the British Library.

This paper meets the requirements of ANSI/NISO Z39.48-1992 (Permanence of Paper).

10 9 8 7 6 5 4 3 2 1

Contents

Acknowledgments

"The frontiers of a book are never clear-cut," wrote Michel Foucault in *The Archaeology of Knowledge.* "Beyond the title, the first lines, and the last full-stop, beyond its internal configuration and its autonomous form, it is caught up in a system of references to other books, other texts, other sentences: it is a node within a network." That network also includes people and organizations, funders and employers, libraries and archives. Those other sentences are found in conversations and discussions, conferences and seminars, classrooms and tutorials, emails and tweets. This book is a very small node within a very large network, and could not have been written without help, support, encouragement, and engagement from many others, to all of whom I would like to express my gratitude.

The Unnatural Trade took a little longer to write than planned, delayed in part by the range of commitments experienced by every mid-career academic but, more recently, by the closure of archives and libraries during the COVID-19 pandemic. Initial research took place in 2013–14 and was generously funded by a Leverhulme Trust Research Fellowship under the working title "Slavery and Environmental Consciousness in British Colonial Writing, 1660–1840." In 2016 to 2019, I was able to conduct further research due to my appointment to a Northumbria University Vice-Chancellor's Professorship. I am very grateful both to the Leverhulme Trust and Northumbria University for this generous research support, without which this project, and others alongside it, could not have been completed.

Academic research in the humanities would be impossible without the ceaseless labor of archivists, librarians, editors, and readers. I would like to express my gratitude to the hardworking and ever-helpful staff at numerous libraries, in particular those working in the Rare Books Room at the British Library; the Main Reading Room at the Library of Congress; Northumbria University Library; the Philip Robinson Library, Newcastle University; the National Library of Scotland; the Bodleian Library, University of Oxford; the Library of the Religious Society of Friends at Friends House, London; and Canterbury Cathedral Library. I am very grateful to my editors at Yale University Press, at first Laura Davulis and more recently Adina Popescu, whose help, advice, and patience have been invaluable, while the book has been immeasurably improved by the careful attention of my copyeditor, Robin DuBlanc. I would also like to thank the anonymous readers of the manuscript whose careful and thoughtful comments did much to shape the book as it now appears.

I have been fortunate in having the support and encouragement of friends and colleagues across several academic departments and research networks. I would like to thank all my colleagues in the Department of Humanities at Northumbria University for collectively providing such a supportive and intellectually stimulating environment in which to work. I thank especially those colleagues in the Eighteenth-Century and Romanticism Research Group and the Environmental Humanities Research Group whose thoughts and comments as I presented various early drafts of this book have been enormously helpful. I would also like to thank my former colleagues at Kingston University London, who witnessed and encouraged the genesis of this book before I moved to Northumbria.

This book has been shaped by the many interactions I have had with friends and colleagues in the academic societies and associations on whose councils and committees it has been a privilege to serve in the years during which this book was composed: the Association for the Study of Literature and Environment, UK and Ireland (ASLE-UKI), the British Society for Eighteenth-Century Studies (BSECS), the International Society for Eighteenth-Century Studies (ISECS), and the Linnean Society of London. The interactions and discussions that I have had as part of the British Group of Early American Historians, the Pre-1900 Americanists in the North, and the North-East Forum in Eighteenth-Century and Romantic Studies have also been invaluable. Scholarly societies and networks are the lifeblood of academia, without which our work would be immeasurably poorer.

It would be impossible to personally thank everyone who has helped shape the development of this book, but I would like to give special thanks to those who provided specific information, offered feedback on drafts or spin-off articles, helped with funding bids, invited me to read from the book as it developed, or just set me off in new and fruitful directions with their thoughts and comments: Regulus Allen, Paul Baines, Lisa Berglund, Kevin Berland, George Boulukos, Conrad Brunström, Jonathan Clark, Katherine Clark, Penelope Corfield, Sue Edney, Markman Ellis, James Fisher, John Gilmore, Ryan Hanley, Christopher Iannini, Adeline Johns-Putra, Richard Kerridge, Peter Kitson, Jessica Moody, Anders Mortensen, Geoff Plank, Sarah Sanders, Tess Somervell, Laura Stevens, and Srividhya Swaminathan.

Finally, more than anyone, I thank my wife Rosie Paice, who has engaged with every page of this book during its long development and who offered her constant support for this project while simultaneously writing her own fabulous book in half the time and with a quarter of the fuss.

Introduction

In May 1789, William Wilberforce (1759–1833) delivered a speech in the House of Commons that was widely hailed as one of the most eloquent pieces of political rhetoric ever heard in Parliament. His topic was the slave trade, which he urged Parliament to outlaw with immediate effect. His evidence was as wide ranging as it was compelling. From Britain to Africa to the Caribbean, the slave trade had had the consequence "of carrying misery, devastation, and ruin wherever its baneful influence has extended." He could not, he told the assembled members of Parliament, "help distrusting the arguments of those, who insisted that the plundering of Africa was necessary for the cultivation of the West-Indies. I could not believe that the same Being who forbids rapine and bloodshed, had made rapine and bloodshed necessary to the well-being of any part of his universe." Wilberforce had recently embraced evangelical Christianity. For him, as for many others in the eighteenth century, the natural world was synonymous with God's creation, and the law of God was also the law of nature. The slave trade, which unambiguously contravened the biblical commandments that outlawed theft and murder, ran contrary to nature, whose law must be consistent across God's universe. Parliament had, therefore, a moral imperative to harmonize statute law with natural law and abolish the trade. This would restore the natural balance of nations. The slave trade "has necessarily a tendency," Wilberforce argued, "to encourage acts of oppression, violence, and fraud, and to obstruct the natural course of civilization and improvements" in Africa. These improvements, he reasoned, would arise naturally from the resources of the African environment and the productivity of its land. "The continent of Africa," he noted, "furnishes several valuable articles of commerce highly important to the trade and manufactures of this Kingdom, and which are in great measure peculiar to that quarter of the globe; and . . . the soil and climate have been found, by experience, well adapted to the production of other articles." Abolishing the unnatural slave trade and replacing it with a trade in Africa's natural resources based on "true commercial principles" would not only "make reparation to Africa, so far as we can" but might reasonably be expected to promote "the progress of civilization and improvement on that continent."[1]

Wilberforce's sense that the slave trade impeded the natural course of commerce and development in Africa was widely shared in the debate about the slave trade that convulsed British politics between 1787 and 1793. In a book published just two weeks before Wilberforce's speech, Olaudah Equiano (c. 1745–97) described Africa as "a country where nature is prodigal of her favours" and argued that "a commercial intercourse with Africa opens an inexhaustible source of wealth to the manufacturing interests of Great Britain . . . the bowels and surface of Africa, abound in valuable and useful returns; the hidden treasures of centuries will be brought to light and into circulation. Industry, enterprize, and mining, will have their full scope, proportionably as they civilize." For Equiano, "the manufacturing interest and the general interests are synonymous. The abolition of slavery would be in reality an universal good." The same argument was made repeatedly by Thomas Clarkson (1760–1846), who carried around a large chest crammed with samples of African natural and manufactured goods to local abolitionist meetings to demonstrate not only the diversity and sophistication of African cultures but also what free and fair trade with Africa might look like. Meanwhile, across the nation, the trade in enslaved Africans was repeatedly condemned as unnatural. In Hull, Wilberforce's hometown, the Reverend John Beatson (1743–98) gave a sermon on the Parable of the Good Samaritan to illustrate "the cruelty of this unnatural commerce." In Dudley, Warwickshire, the Reverend Luke Booker (1762–1835) patriotically argued that "should the shameful and unnatural Commerce be abolished from our Islands . . . *Our* Light, so shining before *other* Nations, they shall see,—shall *imitate* our GOOD WORK." In Edinburgh, George Stewart (fl. 1787) excoriated the environmental despoliation the slave trade produced but predicted that it was naturally unsustainable and "must soon exhaust itself" through depopulation and desertification. "These annual drains," he posited, "have depopulated the shores, rendered the coasts a desert; and there being no supply without penetrating hundreds of miles in to the interior parts of the country, that unnatural commerce seems to be threatened with inevitable dissolution."[2]

Others took the theme of perverted nature even further. Poets saw the trade as unnatural, monstrous, and horrific. "But say, whence first th'unnatural trade arose?" asked William Roscoe (1753–1831), who donated the profits of his 1787–88 poem *The Wrongs of Africa* to the Society for Effecting the Abolition of the African Slave Trade. "And what the strong inducement, that could tempt / Such dread perversion?" Hannah More (1745–1843) offered a similar rhetorical question: "What wrongs, what injuries does Oppression plead / To smooth the horror of th'unnatural deed?" She found none. Instead, she recommended fair trade in Africa's natural resources: "Gold, better gain'd, by what their ripening sky, / Their fertile fields, their arts and mines supply." Alongside poets, pamphleteers proclaimed resistance to slavery as a natural consequence of its horrific nature. In 1791, on hearing news of the revolution of the enslaved population of Santo Domingo, Percival Stockdale (1736–1811), defended "the right which the Negroes inherit, from nature, and from Heaven; from the laws of God, and man, of rising against their oppressours, with a just, and destructive indignation." Such attitudes put Stockdale on the radical wing of the abolitionist movement, which had lost much public support after the news from Santo Domingo reached Britain. He concluded that the uprising was itself proof of both the humanity of Africans and the barbarity of slavery: "If the Negroes were as invariably passive in their horrible state of slavery, as our West-Indians . . . would *have* them to be," he argued, "they might, on some foundation, class them with the lowest, and basest of the animal creation." But to be human is to resist slavery: "The cries, therefore, of the Africans, against their tyrants, is the *Voice*—their revenge, is the *Act*—of NATURE, and of GOD." Even those who thought that gradual abolition would be safer than revolutionary change saw the situation as a perversion of the natural order. "Let servitude be abolished," proclaimed Joseph Priestley (1733–1804), "and leave it to the ingenuity and industry of our countrymen to find a substitute for it. When things are brought into a complex and unnatural state, it is not easy to revert to that which is proper and natural: but in time it will be done."[3]

Whether as impulsive radicals or as cautious reformers, as politicians or as poets, abolitionists of the late eighteenth century agreed that the slave trade was unnatural. To be sure, this was not their only objection. The slave trade was manifestly bloody and cruel. It was sinful and promoted sinfulness. It was unjust and contrary to British notions of liberty. These were major arguments in the abolitionist arsenal, but, alongside these, the notion that the slave trade and slavery itself were "dread perversion[s]" of nature had also become a consistent feature of antislavery discourse by the late eighteenth century, both as a line of reasoning and as a figure of speech. This book asks, therefore, how and why the British slave trade and British colonial slavery came to be understood as unnatural. It shows that from the seventeenth century onward, environmental writers such as naturalists, travelers, and agriculturalists amassed an archive of representations of plantation slavery and the slave trade that late eighteenth-century abolitionists could and did draw upon as evidence for their cause. Many of these earlier writers had represented slavery as a natural element of African society, now naturalized to the New World, a position against which abolitionists recoiled. Part 1 of this book shows how this archive was created, part 2 how it was appropriated and deployed by British abolitionist writers in the last part of the eighteenth century. By observing that almost all descriptions of the slave trade and plantation slavery readily available to British readers before 1760 were written either by naturalists or by travelers with a strong interest in natural history, we can understand the origins of abolitionism to be as much as an issue of eighteenth-century science as one of economics or humanitarian sensibility. Likewise, by considering abolitionists' use of plantation management manuals, another form of environmental writing, we see the abolition movement not only as concerned with the abstract philosophical concept of human rights, but also, and more practically, as a movement for social justice in agriculture. If the antislavery movement, like other movements for freedom and self-determination, was about liberty and the pursuit of happiness, then it was also about life, and the emerging life sciences were at its core.

Plantation slavery and the slave trade that facilitated it were the most egregious of the many abuses that imperialists foisted on their unwilling colonies, but they were part of the wider colonial project. I accept the now widely held view that empire is as much a discursive feature as a strategic one: a set of attitudes and assumptions that must be repeatedly practiced and performed as much as a portfolio of land, cities, oceans, and peoples to be conquered and controlled. But I also see empire as the attempt not only to rob nations of their natural and human resources but also to impose and enforce alien relationships with ecosystems, to introduce both organisms and forms of land management that are at odds with natural landscapes. This combination of imposed cultural attitudes with introduced environmental practices we might call an "unnatural empire," of which the "unnatural trade" was only one part. The diverse authors examined in this book all contributed to the growth of this empire. Some, like planters Richard Ligon (c. 1585–1662) and Samuel Martin (1694–1776), quite literally imposed an unnatural empire with their bare hands, planting, hoeing, and harvesting introduced sugar in Britain's newly acquired Caribbean colonies. Others, like scientists Hans Sloane (1660–1753) and Griffith Hughes (1707—c. 1758), extended its boundaries by adding to its store of botanical, zoological, and geological knowledge. This unnatural empire was merely in its infancy at the start of the nineteenth century. It existed, and indeed continues to exist, as a discursive structure, as a social structure, and as a biological structure. As discourse, it encompasses a set of ideas, attitudes, texts, and utterances that interrogate, explain, and seek to control people, places, and organisms. As a social structure, it imposes hierarchies of power and dominance that maintain systems and networks of consumption and deprivation. As a biological structure, it thrives on transplanted species, exploited organisms, and disrupted ecosystems. The current environmental crisis has many of its roots in the empires of former centuries. Understanding the entangled origins of environmental consciousness, natural history, and antislavery as they developed together in the seventeenth and eighteenth centuries is therefore an important intellectual endeavor for critics, historians, and environmentalists—and becomes essential if we are to continue to understand and address environmental and humanitarian concerns into an uncertain future. Studying the campaign for the abolition of the slave trade, that "dread perversion of nature," offers us an insight into the one of the worst abuses of empire—but also illuminates how the legacies of those abuses can be challenged.

The reasons for the change in public attitudes toward slavery from the mid- to late eighteenth century onward are complex and contested and may never be fully understood. Academic opinion has long been divided between those who see British antislavery as a cynical metropolitan response to dwindling colonial profits and those who proclaim it as a triumph of a new and enlightened humanitarian ethos invigorating European culture. Simple interpretations such as these are not particularly helpful. The growth of antislavery movements undoubtedly resulted from a complex network of interrelated economic, social, and cultural phenomena that came together at the end of the eighteenth century and developed into the nineteenth. The rise of industry, capital, and economic diversification, cultures of sensibility and humanitarianism, the increasing role of antislavery evangelicals, Methodists, and Quakers in public life, and activism and resistance by enslaved people themselves all contributed to the growth of antislavery as a political force. British antislavery also developed in dialogue with intellectual and religious movements in the colonies and across Europe more broadly.[4]

These factors continue to garner considerable academic and public attention and have become the subject of a body of literature too extensive to review here. This book has the more focused aim of contributing to the small but growing subset of studies that consider slavery, culture, and environment in tandem. This literature emerges from the wider field of colonial environmental history, which arguably originates with Alfred W. Crosby's groundbreaking 1972 study *The Columbian Exchange: Biological and Cultural Consequences of 1492.* Crosby's key observation was that, following the European discovery of America, not only people but also a vast array of previously only distantly related organisms crossed the Atlantic in both directions, with profound implications for both American and European ecosystems. A decade later, in *Ecological Imperialism: The Biological Expansion of Europe, 900–1900,* he expanded his thesis to cover a thousand years of global history charting the almost universally deleterious effect of European colonization on non-European habitats. This observation was considerably expanded and complicated by Richard H. Grove, whose 1995 study *Green Imperialism: Colonial Expansion, Tropical Island Edens and the Origins of Environmentalism, 1600–1860* argued that the roots of modern conservationism can be found in the European encounter with the tropics. In part this arose from Europeans becoming increasingly aware of the environmental damage they were causing and in part through cultural engagements with local people and their understandings of the natural world. Grove applies this thesis to the study of colonies across the world, but he says little about the slave trade. Judith A. Carney and Richard Nicholas Rosomoff, by contrast, pay minute attention to the plants imported, tended, and consumed by enslaved Africans in the Caribbean. *In the Shadow of Slavery: Africa's Botanical Legacy in the Atlantic World* (2009) recovers both the cultural and the ecological impacts of this hitherto overlooked aspect of the Columbian exchange. In *Mosquito Empires: Ecology and War in the Greater Caribbean, 1640–1914* (2010), J. R. McNeill extends this investigation to microorganisms, showing how diseases such as malaria and yellow fever impacted colonial history, both slowing colonization and impeding counterrevolutionary forces when colonies asserted their independence. Nevertheless, the Caribbean environment continued to be extraordinarily valuable to Europeans colonists, not least for the riches that flowed from its sugar plantations, whose ecology, agronomy, and culture have been lucidly explored in Sidney Mintz's *Sweetness and Power: The Place of Sugar in Modern History* (1985), Elizabeth Abbot's *Sugar: A Bittersweet History* (2009), and James Walvin's *Sugar: The World Corrupted, from Slavery to Obesity* (2017). As well as introduced sugar, native Caribbean plants were also a source of great interest, and great profit, for their medical use as remedies and cures, as Londa Schiebinger has shown in *Plants and Empire: Colonial Bioprospecting in the Atlantic World* (2004) and, with more emphasis on the knowledge of Africans in the region, in *Secret Cures of Slaves: People, Plants, and Medicine in the Eighteenth-Century Atlantic World* (2017). In *Captivity's Collections: Science, Natural History, and the British Atlantic Slave Trade* (2023), Kathleen S. Murphy shows how specimens gathered by and for naturalists along the routes of the slave trade contributed substantially both to scientific knowledge and to museum collections and herbaria. Eric Herschthal in *The Science of Abolition: How Slaveholders Became the Enemies of Progress* (2021) looks at the equation from the opposite angle, showing how abolitionists from the late eighteenth into the nineteenth century worked to associate scientific progress, in natural history in particular, with freedom and to cast slavery and slaveholders as the enemy of progress.

Literary critics have until recently been less forward than historians in exploring the relationships between culture, slavery, and environment, particularly in the Atlantic and Caribbean sphere, despite numerous critical studies of writing about slavery. This may in part reflect David Fairer's mordant assessment that "ecological criticism has been finding it difficult to gain a purchase on the eighteenth century that is anything other than negative," not least because eighteenth-century literature, unlike the Romantic-era writing that followed it, seemed to earlier critics to be peculiarly abstracted from nature. This, as Fairer shows, is a misreading of the period's writing, but it is nevertheless true that scholars of eighteenth-century literature are only now overcoming their aversion to ecocriticism. Nevertheless, there are some significant works that explore literature and environment in their colonial context. Many of the terms and concerns of literary criticism in this area derive from Mary Louise Pratt's *Imperial Eyes: Travel Writing and Transculturation* (1992), for example, although it is not directly concerned with either slavery or the Caribbean. Susan Scott Parrish's *American Curiosity: Cultures of Natural History in the Colonial British Atlantic World* (2006) shows how African, European, and Native American cultures, driven by curiosity about the organisms and environments of the Americas, combined to create modern European ways of knowing.

Thus, argues Parrish, the discipline of natural history was not merely a European mode of viewing the world, and especially the New World, from the outside, but was instead very largely a colonial discourse, generated by the complex conditions of the Americas and profoundly altering European intellectual culture. Susan Dwyer Amussen pays close attention to the writing of seventeenth-century travelers and natural historians in *Caribbean Exchanges: Slavery and the Transformation of English Society, 1640–1700* (2007). Her contention that "owning people did not come 'naturally' to the seventeenth-century English: slaveholders had to learn to do it" underpins much of my thinking in the early chapters of this book. Christopher P. Iannini's *Fatal Revolutions: Natural History, West Indian Slavery, and the Routes of American Literature* (2012) argues that the growth of the new scientific discipline of natural history is inextricably linked with the development of the Caribbean plantation. He initially pursues this thesis through a careful reading of Hans Sloane's *Voyage to Jamaica* (1707–25), which he sees as a model for all future natural histories of the New World, not least in its depiction of slaveholder violence, although the book goes on to focus on the literature of the early American republic rather than either British abolitionism or natural history in the British Empire. And in *Slavery and the Politics of Place: Representing the Colonial Caribbean, 1770–1833* (2014), Elizabeth A. Bohls reads natural history alongside many geographies and travelers' accounts of the Atlantic world to argue that "the politics of slavery" were effectively "the politics of place" in the abolitionist period.[5]

An important strand of this thinking, which is still unfolding, is the development of critical methodologies that foreground the experience of Black and indigenous peoples in colonized environments. For example, Monique Allewaert's *Ariel's Ecologies: Plantations, Personhood, and Colonialism in the American Tropics* (2013), offers a sophisticated reading of the idea of personhood as it emerged from plantation societies and ecologies in the New World, paying particular attention to the experience of African people through readings of a wide range of text, image, and artifact. Allewaert explicitly aligns her work with "the emerging subfield of postcolonial ecocriticism, which emphasizes that subaltern persons, far from being unacquainted with environmental concerns and thus in need of being integrated into the burgeoning environmental movement of the twenty-first century, have long been among those most directly impacted by the environmental crises that accelerated with the development of colonial capitalism." Postcolonial ecocriticism takes many forms, which is unsurprising for a field that considers writing and environment at local, regional, and global scales across six centuries of European colonization. Of particular importance to the study of the Caribbean region has been the emergence of "Black ecologies" which, argue Justin Hosbey, Hilda Lloréns, and J. T. Roane, consider "the ecological consequences of slavery and its afterlives in the enduring regime of extractivism and disposability shaping Black communities in the Diaspora," but which also challenge "hegemonic modes of environmentalist inquiry" that "tend to efface the critical insights, world-making, and world-sustaining practices of global Black communities." Malcom Ferdinand, in *Decolonial Ecology: Thinking from the Caribbean World* (2021) describes a "double fracture" in which "by leaving aside the colonial question, ecologists and green activists overlook the fact that both historical colonization and contemporary structural racism are at the center of destructive ways of inhabiting the Earth." At the same time, "leaving aside the environmental and animal questions, antiracist and postcolonial movements miss the forms of violence that exacerbate the domination of the enslaved, the colonized, and racialized women." One way to overcome this fracture is to alter the terms of the debate. For example, Ferdinand joins those who challenge the (itself contested) notion of the "Anthropocene," the proposed but, at the time of writing, not yet officially adopted geological epoch during which human activities such as nuclear testing and carbon dioxide emissions can be shown to have left a permanent record in the Earth's geology. For Ferdinand and others, the "Anthropos" element of "Anthropocene" is problematic since human beings are not implicated equally in the origins of the current environmental emergency. Instead, they suggest the term "Plantationocene" to emphasize that the plantation is "at the center of the colonial inhabitation of the Earth. . . . A violent, patriarchal, and misogynistic system, the forced transformation of the Caribbean islands translates into massive environmental destruction, a true 'biological revolution' that overturned the pre-1492 ecosystems."[6]

The Unnatural Trade contributes a strand to this current rapid development of postcolonial environmental humanities, and to some extent bridges Ferdinand's "double fracture" by charting the influence of writing about Caribbean plantations and African environments on abolitionist discourse. "Environmental writing," as I use it here, is an umbrella term that includes natural history writing, agricultural writing, travel writing, and literary representations and celebrations of places and organisms. Academic study of these literary fields is both extensive and accelerating, spurred on by new methods in ecocriticism as well as growing collaboration between scholars across both the humanities and the sciences, and researchers are increasingly aware that the various types of environmental writing are not only defined by their field of scientific inquiry but are also literary genres with their own formal and rhetorical conventions. Natural history in particular is now understood and examined as a literary genre that reflects cultural understandings and engagements with both nature and other nature writing as much as a scientific discipline that attempts to impartially describe the natural world. One objective of such studies has been the attempt to understand what authors mean when they speak of nature. There is no simple answer to this. Raymond Williams famously noted that "nature is perhaps the most complex word in the language." He attempted to resolve this complexity into "three areas of meaning: (i) the essential quality and character *of* something; (ii) the inherent force which directs either the world or human beings or both; (iii) the material world itself, taken as including or not including human beings." The first of these is perhaps the most frequent usage in the period covered by this book. Substances are described as having a solid nature, a slippery nature, or a foul-smelling nature. People have an honest nature or a careless nature. It is the nature of philosophy to raise complex questions or the nature of politics to divide opinion. The slave trade is of a cruel and barbarous nature. Very often in such formulations, the word *nature* is merely used decoratively. One might as well say, and with greater economy, that a person is good, or that slavery is cruel. Sometimes, however, the word indicates that a deeper investigation is taking place, sometimes into the material world itself, including human beings. In this book, the inquiries signaled by the subtly different phrases "human nature" and the "nature of humanity" often strike at the heart of the slavery debate. As well as the debate over whether it was in the uncorrupted nature of human beings that one might justifiably enslave another, there was also an increasingly insistent debate in this period over whether certain peoples, Africans in particular, were in fact human or were of another nature entirely. The attempt to understand the nature of humanity that gave rise to so much of the modern life sciences also led to pseudoscientific racial ideologies that had far-reaching social implications, not least for the future of slavery and abolition.[7]

The second area of meaning that Williams identified was "the inherent force which directs either the world or human beings or both." In this period, in the Christian world, that "inherent force" was almost always identified with God, either as an active force working in a mysterious way or, more distantly, as a first mover who determined the laws of the cosmos and then retired to the supernatural realm. Williams likens this more distant God to a constitutional monarch, with nature "in effect personified as a constitutional lawyer." This is often the primary mode in which natural history and natural philosophy are cast. Abolitionists, like many others in the eighteenth century, also often talk about nature in this way and it is common to see them invoke ideas such as "the laws of nature and humanity" with few metaphysical implications. Many abolitionists were, however, members of congregations that had a less "constitutional" conception of God. Evangelicals, Methodists, and Quakers were all at the forefront of antislavery organizations. At the same time, many natural historians were themselves active Christians, creating works of physico-theology, later renamed natural theology, that sought to understand and glorify the creator through the study of the creation. On closer examination, the so-called age of reason often turns out to be an age of faith, marked by evangelical revivals and great awakenings. "Nature" in much of this writing means "God," the "inherent force" who determined not only the physical nature of the universe but also its moral laws. Indeed, trading in and keeping slaves was increasingly viewed as sinful in this period. Some found scriptural reasons to show why slaveholding was sinful per se. Others argued that slavery simply could not exist without other forms of sin, such as murder, theft, and adultery. In both cases, slavery could be shown to be unnatural since it corrupted, or contravened, the natural law of God.

Those who saw God as a less active force in the world might also see slavery as unnatural. If human beings were naturally altruistic, then enslaving others, alongside other criminal and antisocial behaviors, was a corruption of human nature. This was especially apparent to followers of the sentimental or moral sense philosophers, who argued that natural laws could be inferred from the fact, as they perceived it, that all human beings experienced the world through the same set of senses, and thus experienced pleasure and pain alike. This commonality of feeling was "the inherent force" that gave rise to natural law. On the other hand, there were those who believed, with Thomas Hobbes (1588–1679), that human beings were naturally bellicose and that the presocial state of nature was a war of all against all. If slavery was, as John Locke (1632–1704) maintained, "the state of war continued, between a lawful conqueror and a captive," then it followed that it was a natural phenomenon. Those who accepted Hobbes's dictum that the life of the unsubjected person was "nasty, brutish, and short" could argue that a life in slavery was at least an improvement on a life in the state of nature—which some presumed existed in Africa. In reality, positions were often complex and multifaceted. When abolitionists discuss natural law, natural justice, or natural rights, they sometimes do so by appealing to legal precedent, sometimes with recourse to Enlightenment rationalism and/or moral sense philosophy, and sometimes with reference to God's moral code as revealed in scripture. In an extended treatise on natural law published in 1777, for example, the lawyer, theologian, and veteran abolitionist Granville Sharp (1735–1813) did all three. Sharp noted that "the Jewish constitutions were not strictly consistent *with the Law of Nature in all points*," and could not therefore be used to justify slavery, but cited the sixth-century *Corpus juris civilis*, or Code of Justinian, to remind his readers that "all the best writers, both ancient and modern, agree in adopting that maxim of the Civil Institutes, which declares *involuntary servitude, or slavery, to be 'contrary to the Law of Nature.'* " The Code of Justinian certainly did declare slavery to be unnatural, although how relevant that was to eighteenth-century English law was moot. Most of Sharp's argument, therefore, rests on the sentimental principle that God has provided all animals and humans with "natural affection," from which natural law arises, but has also endowed human beings with reason, which allows them to study the revealed religion, or scripture, which sets out God's laws. Taken together, reason and affection, scripture and nature reveal that slavery, which contravenes all of them, "is the most unnatural Tyranny." Despite the wrangling of philosophers, lawyers, and divines, however, there was in everyday politics no consensus about the nature of slavery. Instead, the eighteenth century saw increasing polarization between those, ultimately the majority, who saw slavery as unnatural and the minority that saw it as part of the natural order. By the end of the eighteenth century, in the British Empire at least, the dwindling latter group was almost entirely composed of people who directly or indirectly profited from the proceeds of slavery.[8]

Williams's third area is "the material world itself, taken as including or not including human beings." This is the main interest of this book. *The Unnatural Trade* is concerned with the discussion of slavery and the slave trade in the diversifying and expanding modes of describing and understanding the material world in the seventeenth and eighteenth centuries that were known to contemporaries as natural philosophy and natural history. In general, natural philosophy concerned itself with both observation and experiment in the search for immutable laws that could describe and predict the behavior of the material world. These are today largely the laws of physics and chemistry. Natural history, by contrast, was more often confined to observation only, describing and cataloguing the animal, vegetable, and mineral species that comprised the material world. These areas are largely the province of zoology, botany, and geology. Even in the early modern period, these distinctions were not quite precise. As more was learned about the physical and chemical composition of both the material and the living world, the laws of physics and chemistry became more central to the study of organisms and rocks. Nevertheless, descriptive natural history was an important and recognized scientific procedure in the seventeenth and eighteenth centuries as well as a sophisticated literary form with its own set of conventions that had been established over a long history.

Although medieval herbals and bestiaries often conflated the two, what we today call the life sciences had been divided into medicine and natural history since ancient times. Both were widely practiced in the classical world, and widely written about, but the towering figures of written natural history were Aristotle (384–322 BCE), Theophrastus (c. 371—c. 287 BCE), Pedanius Dioscorides (c. 40–90), and Pliny the Elder (23–79), from whom many of the conventions of later natural history derived. Aristotle's biological observations are distributed across his entire body of writing, but in particular found in a series of books describing the morphology, taxonomy, anatomy, physiology, and reproduction of animals, of which the *History of Animals* was probably the most influential. Aristotle bequeathed to later natural historians the practice of writing precise and detailed descriptions of organisms based on careful first-hand observation. He was also the first to attempt a serious classification of animals based on morphological principles. His hierarchical organization of organisms into a "scale of nature" would be extremely influential but ultimately highly problematic. The notion of "the great chain of being" had been incorporated into Christian thinking in the medieval period and became the principal system by which nature was understood to be organized. It imagined the entire cosmos as a hierarchy from the least to the most perfect, which in the medieval mind were, respectively, soil and God. Everything else had a precise and fixed position on the chain, with rocks above soil, metals above rocks, plants above metals, animals above plants, and so on. Human beings were at the center; their bodies occupying the highest rung of the physical chain and their souls the lowest rung of the metaphysical chain, which extended through the angelic choirs to God. Although increasingly modified or challenged by eighteenth-century naturalists, this conception of the cosmos would be seized upon by racial theorists such as Edward Long (1734–1813) eager to rank the varieties of humanity or even exclude some peoples from their definition of humanity. In Aristotle's work, nonetheless, it was an important step in the development of the discipline of taxonomy. Aristotle's method was more immediately influential as well. His disciple Theophrastus adopted his method of observation and description, although less so his scale of nature, and applied the method to plants, whereas Aristotle had been primarily interested in animals. *The History of Plants* took its place alongside *The History of Animals*, the two recognized as the foundation stones of botany and zoology.

The other major works of classical natural history to exert influence over naturalists in this period were those by Dioscorides and Pliny the Elder. Dioscorides' *De materia medica* was primarily an herbal: a guide to medicinal plants and their properties, or "virtues." The book was organized primarily by medicinal use rather than by any biological system and contained separate books on aromatics, fats, roots, and wines. Despite the fact that it described only a Mediterranean flora, often very obscurely, *De materia medica* underpinned almost every European botanical work produced before the seventeenth century, and its influence can still be traced in the eighteenth, particularly in the work of the many naturalists who had originally trained as doctors. The most profound legacy of Pliny's encyclopedic *Naturalis historia*, or *Natural History*, was its structure rather than its sprawling and eclectic contents. Pliny ordered his information largely on the basis of scale, starting with astronomy, meteorology, and geography before describing human societies and bodies. His sections on ethnography and human physiology were followed by those on zoology and botany, which also included agriculture. The final volumes of the *Natural History* were more diverse, and included descriptions of minerals and metals, both in nature and in art, which led on to a more general discussion of art, sculpture, and jewelry. Although in later works of natural history the order was sometimes modified, and the art history sections usually omitted, this basic structure underlay many works of natural history in the seventeenth and eighteenth centuries. Others, however, were more influenced by the Aristotelian scale of nature or its derivative chain of being. Natural histories based on this model would start with descriptions of the soil, seen as the link of the chain furthest from God and thus least perfect, and work up through plants and animals. There were numerous hybrid models and, depending on the region or habitat under investigation, the order might vary considerably. Whatever the structure, natural histories in this period often paid far more attention to geology, meteorology, ethnography, and agriculture than they do in our period. In addition, because it was generally believed that God had created the world for human use, the distinction between nature and agriculture was often more a practical than a philosophical distinction. Much inquiry, in the continuing spirit of the medieval herbal or bestiary, was simply to determine for what use God had created plants and animals. In the era of colonial slavery, readers of Caribbean natural histories interested in the appearance, culture, and labor of enslaved Africans could therefore turn to the sections on ethnography, usually toward the start of the book, and on agriculture, generally toward the end.[9]

Colonial naturalists did not, of course, rely exclusively on classical authors. They were also avid readers of contemporary natural history. This was a burgeoning field in full transition from the medical herbalism of the Middle Ages to the systematic biology of the eighteenth century, most famously exemplified in the taxonomic work of Carl Linnaeus (1707–78). While the body of writing concerned with the natural history of the Atlantic world, from Africa to the Americas, is considerably less extensive than the totality of natural history writing from this period, it still consists of many hundreds of primary works. At the same time, there are thousands of primary texts concerned with slavery and the slave trade, and a very significant number of these deal with both slavery and natural history. Rather than attempting to survey all of these, this book reads a selection of the most important or representative published texts, with an emphasis on those that are in dialogue with one another. This means that many important published texts have been omitted because they had limited influence on later authors or, particularly with later texts, do not reveal much about their sources and influences. I have discussed the work of only two seventeenth-century visitors to the Caribbean, for example, Richard Ligon and Hans Sloane, because their work demonstrably had impact on later writers. Those omitted include, for instance, Richard Blome (1635–1705) and the English translator of Charles de Rochefort (1605–83), both of whose Caribbean natural histories reveal contemporary attitudes to enslavement alongside their natural history. Thomas Tryon (1634–1703), who catalogued Caribbean fruits and vegetables, described environmental degradation in Barbados, and issued one of the first antislavery pamphlets in English, has also been excluded, not because his works lack significance in their own right but because there is little evidence that they influenced later abolitionist writers. Likewise, among others, William Smith's (c.1700–1749) *Natural History of Nevis* (1745) and Patrick Browne's (1720–90) *A Civil and Natural History of Jamaica* (1756) offer the historian a wealth of information about both the flora and fauna of those islands and their authors' attitudes toward slavery, but neither appears to have been consulted widely by abolitionists. At the other end of the period under consideration, I have omitted discussion of the monumental poem *The Botanic Garden* (1792) by Erasmus Darwin (1731–1802) and the important *Narrative of a Five Years Expedition against the Revolted Negroes of Surinam* (1796) by John Gabriel Stedman (1744–97). Both intertwine natural history and antislavery sentiment, but both are also curious outliers from the mainstream of abolitionist discourse. Throughout this book, I have concentrated on published work rather than manuscripts because, while manuscripts did circulate, their reach and influence was generally less than published texts and their impact on public discourse is more difficult to evaluate.[10]

Some omissions have been more from necessity than from choice. Much of this study is regrettably but unavoidably focused on male writers. There are many women naturalists in this period, including such celebrated figures as Mary Somerset, Duchess of Beaufort (1630–1715), who was an important botanist and a pioneering gardener who collected and grew plants from around the world including the West Indies. She did not travel, however, nor did she write for publication. Like most women naturalists in this period, she received materials from travelers and correspondents and thus came to know and describe plants and other organisms from samples but did not have the opportunity to observe the physical or the social environment from which they were taken. Many more women published antislavery literature, including Hannah More (1745–1843), whose 1788 *Slavery, a Poem* is considered in detail, but surprisingly few of them, More excepted, engaged with ideas about slavery and nature. It has also been difficult to locate many discussions of slavery in relation to ideas of nature and the natural world in the published works from this period of those who had themselves suffered enslavement. The notable exception is Olaudah Equiano, whose 1789 *Interesting Narrative* is discussed in depth. Otherwise, the publications of Black writers in this period tend to reveal more about the social, political, and spiritual concerns of their authors than their ideas about the natural world.

The Unnatural Trade is an interdisciplinary work that considers literary, historical, and biological writing and that aims to be useful both to scholars and to general readers in all those fields and beyond. In the hope of engaging with a broader audience, I have sometimes included contextual or explanatory material that might be self-evident to a specialist reader within a single discipline, but may be less familiar to those from other disciplines. I have, for example, provided brief biographies for all but the most celebrated figures I discuss. Throughout the book, I have consistently applied the scientific convention of providing binomial or Latin names for species, which allows for precise identification, even though this may seem unusual in parts of the book dedicated to, for example, poetic analysis. I have also tried where possible to identify the species being discussed by seventeenth- and eighteenth-century authors, even where that identification is tentative or speculative. Identifying, or attempting to identify, precise species may sometimes seem of secondary importance to readers more interested in textual strategies, but doing so provides important insights into a naturalist's method, mindset, and competence as well as giving vital information about the environment they are describing.

This book makes use of many terms that are complex, contested, or that have changed their meaning across time. I have used these carefully, even though space prohibits a long discussion of etymology and changing usage. I have generally used "African" to describe people both of African birth and African descent and "European" to describe people of European birth and European descent, even though, by the end of the period under discussion, many of the people I discuss would more accurately be described as African American, African Caribbean, or Anglo-American. The terms "African" and "European" are of course themselves vague and imprecise, describing people from diverse and often distant regions, but they were used widely across the seventeenth and eighteenth centuries and continue to be used in the twenty-first. I avoid referring to enslaved people as "slaves," where possible using "enslaved person" or "enslaved people" to emphasize that enslavement was the deliberate act of an enslaver. I use the term "antislavery" to denote the full range of attitudes critical of slavery and the slave trade. "Abolitionism," however, I use more precisely to describe attitudes and actions, generally after the 1770s, that are directly focused on ending them. Terms such as "science" and "scientist" were used very differently, if at all, before the nineteenth century, and while terms describing some scientific disciplines, such as botany and zoology, have a long provenance, others, such as biology or ecology, are more recent. I have used such words cautiously but have not entirely excluded the use of modern terms to describe early modern practices where the meaning is clear. I have also used the term "unnatural" with caution. Although the word was used frequently by abolitionists to describe slavery and the slave trade, beyond abolitionist discourse the term was more often used to denote a rupture in familial relations, as when a parent fails to care for a child, or a child rebels against a parent. From this, it was sometimes extended in this period to describe civil war, including the American War of Independence. The term also had currency as a loose and shifting euphemism for various sexual activities and orientations that in this period might have included, but were not limited to, same-sex relationships, incest, bestiality, interracial sex, and sometimes extramarital sex.

The Unnatural Trade is divided into two parts. Part 1 shows how an archive of descriptions of slavery and the slave trade was created by naturalists and other travelers with an interest in natural history. Part 2 shifts the perspective to demonstrate the ways in which this archive was appropriated and deployed by abolitionist writers from the 1770s onward to show that slavery was unnatural. The argument that the slave trade was a "dread perversion" of nature was only one of many put forward by abolitionists in the last decades of the eighteenth century, but the story of its development and deployment illustrates the ways in which they made use of diverse and sometimes highly specialized literatures to build a case that was both convincing and rhetorically persuasive. Despite the best efforts of planters and slave traders, the British public did not accept that slavery was a natural phenomenon, or the slave trade an unavoidable necessity. Although abolitionists shared in many of the prejudices of their day, and while their focus on African prodigality may inadvertently have hastened public acceptance of imperialism in Africa, eighteenth-century readers became persuaded that both the slave trade and the system of forced labor it serviced were unnatural. This book shows how that came about.

PART ONE

Building the Archive

CHAPTER I

"*The Cord that bindes up all*"

RICHARD LIGON AND THE NATURAL HISTORY OF BARBADOS

A hundred years before William Wilberforce and his generation of abolitionists were born, an exiled Stuart courtier confronted the thorny relationship between Caribbean slavery and the natural environment. Richard Ligon's *True and Exact History of the Island of Barbados,* first published in London in 1657, is principally remembered as the source for the story of Inkle and Yarico, a reputation that almost certainly saved the book from obscurity and that allowed it to have an impact on abolitionist discourse more than a century after its composition. Retold in increasingly sophisticated forms in the eighteenth century, this story of love and betrayal in exotic lands had its origins in just two paragraphs of Ligon's *History,* in which he describes his conversations with "an *Indian* woman" who was "a slave in the house" where he lived in Barbados during the late 1640s. From these discussions he learned how she was betrayed into slavery in Barbados. The woman, whose name is later revealed to be Yarico, was originally from an unspecified location on the American mainland. A small party from the crew of an English ship went ashore at this place, where they were ambushed—most of them were killed. However, "A young man amongst them stragling from the rest, was met by this Indian Maid, who upon the first sight fell in love with him, and hid him close from her Countrymen (the Indians) in a Cave." When the coast was clear, the young man returned to the ship, taking the "Indian Maid" with him. "But the youth, when he came ashoar in the *Barbadoes,* forgot the kindnesse of the poor maid, that had ventured her life for his safety, and sold her for a slave, who was as free born as he: And so poor *Yarico* for her love, lost her liberty." The story became famous in the eighteenth century after being adapted by Richard Steele (1672–1729) in 1711 for the *Spectator.* Steele took Ligon's story, gentrified it, and provided Ligon's young man with a name: Thomas Inkle. In the decades that followed, the tale was adapted in verse and prose, translated into several languages, and even produced as a comic opera. Its popularity is evidence of the appetite in eighteenth-century Britain for tales about moral ambiguity and wrongdoing on the colonial frontier, but also hints at an embryonic, if ill-defined, sense of the wrongfulness of slavery.[1]

Steele explicitly credited "*Ligon's* Account of *Barbadoes*" in his *Spectator* paper, ensuring that the story's creator remained visible to posterity. Today, critics see Ligon's *History* as a crucial moment in the development of what Keith Sandiford has called the "Caribbean-Atlantic imaginary," while historians see it as a vital record of the early stages of the British colonial project and a key moment in the inception of colonial science. As Raymond Stearns has pointed out, "Ligon's thin volume became a starting point for subsequent English studies of the West Indies, and a basis upon which the Royal Society of London formulated various sets of 'Inquiries' directed to its correspondents there." Ecocritics and environmental historians have also begun to take an interest in its record of the impact of Barbadian slavery on the environment. For example, Keith Pluymers argues that "Ligon provided one of the richest accounts for aesthetic and affective environmental thinking in the seventeenth century." By turns a voyage narrative, a natural history, and a plantation management manual, and representing not just Barbados but also the African Cape Verde islands, it was both reprinted and plagiarized in the late seventeenth century, while its author's observations continued to be cited well into the eighteenth. Ligon's *History* provides both the starting point and an enduring source of influence for those later writers who believed slavery was contrary to nature.[2]

Ligon arrived in Barbados at a crucial moment in the island's social, economic, and environmental development: the point at which planters introduced the sugarcane (*Saccharum officinarum*). His book is accordingly a unique and important document of the most significant environmental transformation in the island's history. The *History* also marked a transformation late in the life of its author. Born in the 1580s, he was, as he tells us, more than sixty years old when he set out for the island in June 1647. Little is known of his life before his arrival in Barbados, but the references to aristocratic pursuits such as hawking, hunting, painting, and music that pepper the book, alongside descriptions of the royal family, tell of a life at the top of society, or not far from it. Like many, he lost much in the 1640s. As a Royalist who "spent time at court under Charles I, and may possibly have been a page to Anne of Denmark," he was on the losing side in the Civil War, but he also suffered a financial setback when a fen drainage scheme in Lincolnshire backfired. Destitute and politically isolated, he decided to try his chances in Antigua. Even that scheme went awry. Of the two ships that set out on the expedition in June 1647, the one carrying "men victuals, and all utensell's fitted for a plantation" was lost. Instead, Ligon stayed about three years in Barbados, working as a plantation overseer, before falling seriously ill with a fever, which he cured by taking dried turtle penis in beer. After recovering, he returned to England, where he was thrown into a debtor's prison. In an attempt to turn these unpromising experiences to account, his *History* was composed while he was imprisoned. The book remains eminently readable to this day. Its author comes across as a man of wit and learning whose principal skills were his acute powers of observation, his interest in everything and everyone he encountered, and his lively but precise writing style. Such characteristics make for a fine author, but they are also the hallmarks of a successful naturalist, and it is significant that the first substantial discussion of British colonial slavery takes place in what is also the first substantial natural history of a British Caribbean colony. Ligon's support for the fledgling Barbadian sugar industry, on the one hand, and his ambivalent attitudes toward slavery combined with his holistic and generally benevolent worldview, on the other, created sites of textual uncertainty and reflection that both directly and indirectly influenced the later writers who more explicitly represented slavery as unnatural.[3]

THE VOYAGE TO BARBADOS

Ligon's *History* begins with a voyage. His journey from England to Barbados aboard the 350-ton *Achilles* took several months and included a stay in the Cape Verde Islands, some 650 kilometers (400 miles) from present-day Dakar on the coast of Senegal. Often overlooked by readers eager to reach his account of Barbados, the opening twenty pages of the book might be better seen as the author's preparation for the main event. Writing in 1653, several years after the experiences he describes, Ligon rehearses his skills as natural historian and ethnographer in his account of the transatlantic voyage and West African islands and signals his intention to provide a lucid and accurate description of what he encountered. Avoiding the more direct course taken by modern airliners, the *Achilles* took the longer route, traveling almost due south from

Falmouth to Santiago in Cape Verde and then due west from there to Barbados. As well as this route being safer, allowing the ship to reprovision along the way, mid-Atlantic ocean currents also made it faster. The vessel sailed through the Madeira and Canary archipelagoes without landing, but even from shipboard Ligon offers precise geographical and meteorological observations balanced by inquiry and discussion. Although only early summer, the islands of the Madeira archipelago "which lyeth in 33. degrees to the *Noreward*" were "so miserably burnt with the Sun, as we could perceive no part of it either Hill or Valley, that had the least appearance of green." The aptly named "inconsiderable Ilands called the *Deserts*" (the Ilhas Desertas) were "burnt worse than the other, so that instead of the fresh and lively greenes, other Countreys put on at this time of the yeare: these were apparrel'd with Russets, or at best *Phyliamorts.*" Ligon observes precisely, but he also checks to see if this is normal. The seamen tell him "that they had never seen it so burnt as now, and the *Leeward* part of it was, at other times, exceeding fruitfull and pleasant." Ligon adds the useful information that the islands were, in their account, normally "abounding with all sorts of excellent fruits, Corne, Wine, Oyle, and the best Sugars; with Horses, Cattell, Sheep, Goates, Hogges, Poultrey; of all sorts, and the best sorts of Sea fish." These few lines present in miniature not only Ligon's scientific method but also that of European exploration in Africa. The traveler makes careful observations, seeks local knowledge, and from these reaches a synthesis that sets out the land's utility for European commerce.[4]

Ligon was not the first to follow this pattern, but he is unusual for the detail of his observations, both in Barbados and on the voyage. When the ship passes the next island south from Madeira, it is at such a distance that he can discern little, but he still notes that "the general Landscape of the hills seemed to one very beautifull, gently rising and falling, without Rockes or high precipice." Seeking further information, he imparts the useful knowledge that "this Iland is famous, for excellent Salt, and for Horses, which in one property, excell all that I have ever seene; their hooves being to that degree of hardnesse, and toughnesse, that we ride them at the *Barbados,* downe sharp and steepie Rocks, without shooes." The observation reinforces both Ligon's utilitarian approach to natural history and the island's integration into a transatlantic system of commerce and species exchange. Ten leagues further south he likewise observes "the Ile of *May;* famous for store of excellent Salt." Ligon's information may have been tailored to the needs of European travelers, but it was not always accurate. These two islands, which he locates in the Canary archipelago, he names Bona Vista and May. These, more accurately Boa Vista and

Maio, are in fact two of the three most easterly islands in the Cape Verde archipelago. The lapse is unusual. Ligon's observations are in general as accurate as they are detailed. The error suggests, however, that Ligon is recording his observations from memory rather than from notes or a journal. If this is the case, he had extraordinary powers of recall, a few errors notwithstanding.[5]

The Cape Verde islands, although uninhabited when they were discovered by Portuguese navigators in the 1450s, were rapidly integrated into the Afro-European Atlantic world. After the European discovery of America half a century later, they became an important conduit of trade and, as well as salt and horses, became notable centers of the slave trade. Ligon's expedition took full advantage of Cape Verde's strategic and commercial location but also its distance from European manufacturing centers. The plan was to stop at "St. *Jago,* one of the Ilands of *Cape Verd;* where wee were to trade for *Negros,* Horses, and Cattell; which we were to sell at the *Barbados.*" By way of trade, the ship carried "Cloath, Bayes, Stuffes of several kindes, Linen Cloath, Hats with broad brims, such as Spaniards use to weare, and were made in London purposely to put off there." The pattern is precisely that of the triangular trade in which manufactured goods were exchanged for enslaved people on the coast of Africa who were then taken across the Atlantic in the infamous Middle Passage. The *Achilles,* also carrying colonists, goods, and livestock, was not solely a slave ship but its pattern of trading anticipated the British slave trade that would flourish after 1660.[6]

In the event, trading on Santiago proved less straightforward than anticipated and Ligon's party "found that it was farre better, for a man that had money, goods, or Credit, to purchase a plantation [on Barbados] ready furnisht, and stockt with Servants, Slaves, Horses, Cattle, Assinigoes, Camels, &c." If Ligon's list is also a hierarchy, it shows he ranked indentured and enslaved humans just above horses, an ordering that echoes contemporary notions of the great chain of being, but that also places those people firmly in the legal category of chattel. This is not a point on which he elaborates in his description of Cape Verde. Indeed, his account of its inhabitants is more voyeuristic than ethnographic. He lampoons the grotesque appearance of the governor of Santiago on the one hand while on the other lavishing extravagant and lascivious praise on the beauty of several African women he encounters. The first he calls "a Negro of the greatest beautie and majestie together: that ever I saw in one woman" while the next, two "pretie young Negro Virgins," were "creatures of such shapes, as would have puzzelld *Albert Durer,* the great Mr of Proportion, but to have imitated." These extended passages veer toward the pornographic—although Susan Dwyer Amussen argues that, while "Ligon wrote as a titillated voyeur . . . the courtly language of his description especially suggests an ironic detachment." It was not only the island's women that attracted his gaze. In the "valley of pleasure" in which he encountered two of the women, he also found "woods of pleasant trees" in which he saw "flying divers birds, some one way, some another, of the fairest, and most beautifull colours, that can be imagined in Nature." These are spotted among more than thirty species of useful plants, ranging from "high and loftie trees, as the *Palmeto, Royall, Coco, Cedar, Locust, Masticke, Mangrave,*" and several others to exotic fruits such as "the Plantine, Pine, Bonano, Milon, water Milon, &c. and some few grapes." These are undoubtedly useful, but Ligon admired the plants of the island not only for their utility but also for their beauty and fragrance: "Other kinds of trees, we found good to smell to, as Mirtle, Jesaman, Tamarisk, with a tree somewhat of that bignesse, bearing a very beautifull flower. The first halfe next the stalke, of a deep yellow or gold colour; the other halfe, being the larger, of a rich Scarlet: shap'd like a Carnation, & when the flowers fall off, there grows a Cod, with 7 or 8 seeds in it, divers of which we carried to the *Barbados* and planted there: and they grew abundantly, and they call them there, the St. *Jago* flower, which is a beautifull, but no sweet flower." The description closely matches the peacock flower, *Caesalpinia pulcherrima,* probably originally from East Asia but now widely distributed throughout the tropics. Although no longer called the "St. Jago flower," the plant is sometimes known as Spanish Carnation, sometimes as Barbados Pride, and is today the national flower of Barbados. The passage presents the intriguing possibility, therefore, that Ligon not only wrote the first natural history of Barbados but also introduced its national flower. Whether or not he was indeed the first to take the flower to Barbados, the casual collection and distribution of seeds from one location to another is nonetheless a hallmark of the early modern naturalist. With no understanding of the damage that introduced species can inflict on sensitive island ecosystems, travelers between the Old World and the New both deliberately and inadvertently transformed the flora and fauna of both sides of the Atlantic. Ligon's seemingly whimsical decision to collect and relocate a few seeds of a plant he admired was one small action in the vast ecological transformation that Alfred Crosby has memorably characterized as "the Columbian exchange."[7]

If the island of Santiago was full of wonders, the voyage itself offered plenty of opportunities to observe biological and meteorological phenomena, particularly in the seas around the Madeira, Canary, and Cape Verde archipelagoes. Ligon describes many fish, birds, marine mammals and reptiles, and the Portuguese man o' war (*Physalia physalis*), which he appropriately calls a "carvil" after the Portuguese caravel rather than after the British man of war—although Susan Scott Parrish has noted that the reference to the "carvil" probably also alludes allegorically to imagery of the animal in the " 'Godly Feast' rendered by Desiderius Erasmus in his book of that name, originally printed in 1522 and a staple of seventeenth-century English humanism." This is undoubtedly a significant context, but Ligon's interest in the relationships between the marine species he observes is perhaps more important and would prime him for a holistic reading of the Barbadian natural and social landscape. Natural historians of the period tended to discuss species in isolation. Ligon is more interested in their interactions with each other and their physical environment. While this approach is derived from older organic conceptions of a divinely ordered creation in which all organisms fulfill the role allotted to them by God, Ligon's holistic reading of the environment nevertheless strikes modern readers as intuitively ecological.[8]

This is neatly illustrated in his account of the relationship between flying fish, dolphins, and seabirds. "The Dolphins," says Ligon, "pursue the flying Fish, forcing them to leave their knowne watry Elements, and flye to an unknowne one, where they meet with as mercilesse enemies; for there are birds that attend the rising of those fishes; and if they bee within distance, seldome fayle to make them their owne. These birds, and no other but of their kinde, love to straggle so far from land; so that it may be doubted, whether the sea may not bee counted their naturall home; for wee see them 500 leagues from any land." It is difficult to know precisely which species Ligon is observing. Although Ligon compares the dolphins to "porpisces," the species is almost certainly the common dolphinfish or dorado (*Coryphaena hippurus*), familiar on American menus as "mahi-mahi." There are about sixty species of flying fish, and many hundreds of species of pelagic seabird. The bird is described as "a kinde of sea Hawke, somewhat bigger then a Lanner, and that colour; but of a far freer wing, and of a longer continuance." The description fits Cory's shearwater (*Calonectris borealis*), although the main predator of the dolphinfish is the magnificent frigatebird (*Fregata magnificens*), but while we can never be entirely sure which species Ligon observes, his understanding of the ecology of the sea-surface habitat is striking. He considers both the biotic environment of fish and birds and the abiotic environments of water, land, and air. He describes a complex predator-prey relationship with two predators operating in three dimensions involving both aerial and marine habitats. He locates the habitat in both its local and global contexts and speculates on what might be the "naturall home" of the seabirds. These are all profound observations, prompted by the need to make sense of an unfamiliar habitat. To help, he develops an extended metaphor comparing the interactions he observes at sea with the more familiar scene, for an early modern gentleman, of a hunt in Windsor Forest. With the expert eye of one who was "bred a Faulconer in my youth" he admires the seabirds' aerobatics and notes that "at the times they grow hungry, they attend the *Dolphins,* who are their Spaniels; and where they perceive the water to move, they know they are in Chase, of the flying fish; and being neere them, they rise like Coveys of Partridges by 12 and 16 in a Covey, and flye as far as young Partridges, that are farkers, and in their flight these birds make them their quarry." "Farker," a variant spelling of "forker," an East Anglian dialect word for an unpaired partridge, hints at Ligon's origins, but also reminds us that he was an accomplished huntsman, used to observing animal behavior in the field. The gray partridge (*Perdix perdix*) is well known for its habit of rising suddenly and flying rapidly and noisily for a short distance. In Ligon's mind, the dolphins are performing the role of the hunter's dogs, flushing flying fish out of the sea as spaniels flush partridges from the undergrowth. Once airborne, the partridges are taken by the falconer's hunting bird just as the flying fish are taken by the seabirds. The extended metaphor is of course anthropomorphic. No human huntsman is directing the dolphinfish and seabirds, and the apparent cooperation between them is merely in the observer's imagination, even if to the early modern mind this might have appeared as evidence of divine purpose. Nevertheless, although articulating it in early modern terms, Ligon is clearly grasping that different organisms in different environments can perform similar ecological roles. Even before he arrives at Barbados, therefore, Ligon has primed himself with an awareness of new environments that we might call "proto-ecological," arising in part from his training as a falconer and in part from his close observation of a new habitat while on his transatlantic voyage.[9]

"THAT PERFECTION OF BEAUTY": LIGON AND THE BARBADIAN ENVIRONMENT

Ligon's first impressions of Barbados are likewise framed in proto-ecological terms, leading him to develop a moral-bearing extended metaphor about the government of the state. Approaching from the east, his first sight of the island was probably of the parishes of St. Philip and Christ Church, which he thought "extreamly beautifull." Most visitors today arrive at the airport, also in Christ Church parish. From the air, one's first impression is of a densely populated and highly developed island with housing scattered throughout a neat patchwork of fields—and a surprising absence of trees. Ligon observed the first stages of this development, but what interested him most was the now-vanished forest. He observed that "the high, large, and lofty Trees, with their spreading Branches, and flourishing tops, seem'd to be beholding to the earth and roots, that gave them such plenty of sap for their nourishment, as to grow to that perfection of beauty and largenesse." In turn, "they, in gratitude, return their cool shade, to secure and shelter them from the Suns heat, which, without it, would scorch and drie away. So that bounty and goodnesse in the one, and gratefulnesse in the other, serve to make up this beauty, which otherwise would lie empty & waste." Ligon may have based his description of Barbadian forest hydrology at least in part on Aristotle's *Meteorology,* which described and attempted to explain the relationship between air and water in considerable detail. Nevertheless, his sophisticated account of a self-regulating system made up of interrelating biotic and abiotic elements depicts an ecosystem in all but name. He recognizes that water and nutrients are supplied by the soil, and heat and light are supplied by the sun, but he also understands that this is a system in equilibrium and that tipping the scales will destroy the balance that maintains the whole. Removing the soil's "sap" will kill the trees. On the other hand, killing the trees would dry up the sap. Trees and sap are mutually dependent and, crucially, in harmony without the need of human management.[10]

Ligon also saw in the forest of Barbados a metaphor for the state. Just as the trees gratefully offered their shade to the nourishing earth, so the people of a harmonious state should offer grateful obedience to the rulers who protect them:

> And truly these vegetatives, may teach both the sensible and reasonable Creatures, what it is that makes up wealth, beauty, and all harmony in that *Leviathan,* a well governed Common-wealth: Where the Mighty men, and Rulers of the earth, by their prudent and carefull protection, secure them from harmes; whilst they retribute their paynes, and faithfull obedience, to serve them in all just Commands. And both these, interchangeably and mutually in love, which is the Cord that bindes up all in perfect Harmonie. And where these are wanting, the roots dry, and leaves fall away, and a generall decay, and devastation ensues. Witnesse the woefull experience of these sad times we live in.

Comparing the nation to a garden or other cultivated land is a trope as old as Eden, and common in early modern literature, but Ligon rather surprisingly compares "a well governed Common-wealth" with a wild and self-governing tropical forest rather than a managed English landscape. This is not much in accordance with Thomas Hobbes's conception of the state as an overriding leviathan, an idea that Ligon explicitly invokes (although there is no evidence that he had actually read Hobbes). By contrast, Ligon invokes an organic conception of the state well in accord with his Royalist values and deeply conservative in nature. In the view of Susan Scott Parrish, Ligon "patently avows an essentially hierarchical, aristocratic, Christian humanist worldview within which a reciprocity between the rulers and the ruled ideally existed." Jane Stevenson has argued that this was "an anachronistic ideal of service and loyalty" and that the idea that "it could, might, or should apply to the West Indies is so counter to the islands' subsequent history; the fact that this strain of political thought was ever brought there at all has barely been noticed." Stevenson convincingly shows that much of Ligon's subsequent disquiet about slavery derives from this ideal, but we should also note that it is consistent with an approach that emphasizes natural and reciprocal relationships. Ligon no doubt saw these relationships as divinely ordained, but nevertheless his naturally occurring forest and his ideal commonwealth share important characteristics. The passage is an extended metaphor comparing the affairs of humanity to the processes of nature. We might think of it, indeed, as an environmental fable.[11]

Ligon's admiration for the ungoverned forest makes a political point and is not thereafter sustained. He soon adopts the planter's mentality, seeing woodlands as an obstacle to development and accepting that the colonists' main objective is to clear trees and plant crops. He gives a short historical account of the colony's early decades, which is necessarily secondhand, given his recent arrival, but it is nevertheless noteworthy that he emphasizes the environmental rather than the political history of Barbados. The first years are a struggle to survive in the shadow of an unforgiving rainforest. "Ships were sent, with men, provisions, and working tooles, to cut down the Woods, and clear the ground, so as they might plant provisions to keep them alive." This turned out to be hard work, as did early and failed attempts to grow tobacco, because "the Woods were so thick and most of the Trees so large and massie, as they were not to be falne with so few hands; and when they were laid along, the branches were so thick and boysterous, as required more help, and those strong and active men, to lop and remove them off the ground." Ligon has put his finger on the central problem of the new colony. To succeed, the colonists will need to find the labor required to transform the island from economically unproductive rainforest to lucrative plantation. Their solution, which Ligon does not directly mention in this passage, was to import enslaved laborers from Africa, but at this point in his narrative he instead portrays a heroic, but failing, struggle of European settlers with an unrelenting environment, their goal to find an economic miracle crop that will secure their prosperity. The crop would turn out to be sugar. "When the Canes had been planted three or four years," he tells us, "they found that to be the main Plant, to improve the value of the whole Iland." Great efforts were made to understand and develop sugarcane, and the planters "bent all their endeavors to advance their knowledge in the planting, and making Sugar." Ligon's account would influence future historians, leading to the nineteenth-century notion of a seventeenth-century "sugar revolution" on the island in which sugar was introduced, the plantation system and slavery followed, and the system was then exported throughout the Caribbean. This was long the accepted historical orthodoxy, with an extensive historiography, but some recent scholarship suggests that despite Ligon's firsthand view of this revolutionary moment, a plantation slavery system was in fact already at least partially established and the introduction of sugar merely intensified it.[12]

Ligon was working for a sugar planter and was hoping to sell a book that at least partially concerned the management of a plantation, so it is understandable if his account of the island emphasized the role of sugar. In any case, sugar undoubtedly transformed the island's society and economy, but its effect on the island's ecosystems would also be overwhelming, as recent studies have shown. A pioneering study of the human impact on the Barbadian environment was made by the biogeographer David Watts in the 1960s. Watts based much of his study on archival sources, in particular Ligon's *History* and Griffith Hughes's *Natural History of Barbados,* but he also conducted an ecological study of the small areas of relict precolonial woodland on the island, particularly at Turner's Hall Wood, which is now a nature reserve. Contrasting botanical accounts in the archives with the remaining flora, Watts concluded that before 1627, "Barbados was covered by a well-developed, dense tropical 'forest' which throughout stretched down to the littoral" but which, unlike a true rainforest, was adapted to seasonal periods of aridity. He also suggested that this forest rapidly succumbed to the colonists, noting that "the pre-European vegetation associations were to a very large extent systematically destroyed by man during the first four decades following English colonisation in 1627. In particular, the seasonal rain forest, and the xerophytic [drought tolerant] forest and scrub were quickly removed to make way for commercial crops, more especially cotton, tobacco and sugar cane." Less than half of the island's forest had been cleared by the late 1640s. The introduction of sugar-cane in the 1640s transformed the island's ecology as well as its society and economy, and thus "an extensive final phase of forest destruction commenced after 1650." By 1665, Watts argues, "all but the most isolated patches of forest on steep gully sides, and in the Scotland District, had been cleared." Thereafter, Barbadians were forced to import wood and timber from neighboring islands and from as far away as New England. More recently, however, Keith Pluymers has argued that the deforestation may not have been quite so absolute. Frequent references to woodlands in property records "show that colonists valued woods" as an important local resource and "many had a sophisticated understanding of wooded lands that included multiple, differentiated uses." The New England trade may have been driven by economic rather than environmental pressures, and plantations may have retained stands of trees or wooded boundaries for longer than Watts realized. Nevertheless, whether partial or nearly total, forest clearance substantially reduced the island's biodiversity and opened the way for introduced species. "Today," Watts observes, "most plants are non-native, and of these sugar cane is most abundant, having usually covered about eighty per cent of all land in the island since the sixteen-fifties." This pattern of colonization, clearance, and introduction of non-native species is now depressingly familiar to ecologists, who have witnessed its dramatic and deleterious effect on biodiversity in islands and forests around the world. Ligon was witnessing one of the first instances in human history of the widespread clearance of tropical forest to create a cash-crop monoculture farmed by exploited workers for the profit of transnational agribusiness. Such terms were not available to him, but Ligon nonetheless clearly understood the purpose of the new colony well.[13]

UNDERSTANDING THE SECRETS OF THE WORK: LIGON AS PLANTER

Ligon was the first English writer to offer advice on the cultivation of the sugarcane. The *History* combined voyage narrative and natural history with practical advice to colonists on a range of topics, sugar cultivation included. It also adhered to early modern humanist conventions regarding statecraft and godly behavior. As Susan Scott Parrish reminds us, the book is "not only a practical manual but also a Christian humanist how-to directive for conscientious plantation" even though "Ligon tried—and failed—to reconcile an older Christian humanism and trader cosmopolitanism with the newer forces on the ground in the Atlantic world." His lengthy discussion of sugar comes at the end of the natural history section of his book, immediately following his fulsome and celebrated description of the pineapple, which was an exotic, almost mythical, fruit to seventeenth-century English readers, increasingly associated with royal magnificence, or monarchical excess, depending on your point of view. Ligon described the pineapple and the other plants he had covered in a page or less for each, but his account of "*Sugar Canes, with the manner of planting; of their grouth, time of ripenesse, with the whole process of Sugar-making*" is considerably lengthier and accompanied by several pages of diagrams giving precise dimensions for the construction of an ingenio, or apparatus, for the production of sugar (from the Spanish *ingenio azucarero:* a sugar refinery). Ligon clearly understood that many potential purchasers of his book would be more motivated by the desire to run a profitable plantation than by a thirst for scientific knowledge. Despite this, however, his sugar passage begins with historical rather than practical information as he offers a firsthand account of the problems the planters faced when first introducing the crop. "Finding them to grow," Ligon records, "they planted more and more, as they grew and multiplyed on the place, till they had such a considerable number, as they were worth the while to set up a very small Ingenio, and so make tryall what Sugar could be made upon that soyl. But, the secrets of the work being not well understood, the Sugars they made were very inconsiderable, and little worth, for two or three years." The account illustrates the chancy and experimental nature of sugar production in early Barbados. Although Spanish and Portuguese planters had been producing sugar for many years, long preceding their discovery of the New World, in fact, the English were latecomers to the business. Not knowing if the soil and climate would support the cane, they evidently planted test patches first before going to the expense of building an ingenio, a large apparatus that was among the most complex agro-industrial facilities of the time. Ligon supplies detailed instructions for its construction and operation, no doubt with an eye to the market of planters and would-be planters. His descriptions are technical and precise, and while he speaks very little about the people operating the machinery, there are enough references to "Negroes" to show that from the very start the Barbados sugar industry relied on slave labor both in the fields and in the refinery.[14]

Earlier in the book, following his description of the "Negroes" and "Indians" of Barbados, Ligon had set out an ethnographic account of the "Masters." In his view, these are "men of great abilities and parts" who manage large estates, successfully feeding "two hundred mouths" or more and keeping slaves and servants in good order. Before any of this work can begin, however, "the first work to be considered, is Weeding, for unlesse that be done, all else (and the Planter too) will be undone; and if that be neglected but a little time, it will be a hard matter to recover it again, so fast will the weeds grow there." Ligon's account represents sugarcane cultivation as a battle between planters anxious to bring in a safe harvest and invasive weeds that threaten to overwhelm their crops and their effort, since only "after weeding comes Planting." If Ligon had compared "a well governed Common-wealth" with a wild and self-governing tropical forest rather than a carefully managed English farm, here he enters more familiar metaphorical territory: in the safe hands of the planters, Barbados is emphatically a weeded garden that has not gone to seed. It is significant, however, that he insists that "this work of planting and weeding, the Master himselfe is to see done; unlesse he have a very trusty and able Overseer; and without such a one, he will have too much to do." Ligon himself had been an overseer on Sir Thomas Modyford's plantation, presumably seeing himself as "trusty and able," and so this may be a piece of self-congratulation on his part. It might also be seen as veiled Royalist sentiment, implying that only the head of the nation, metaphorically represented by the head of the plantation, or a trusted lieutenant could safely direct the removal of noxious components of the state and replace them with more productive elements. If this is so, Ligon does not labor the point—he was, after all, writing as a Royalist anxious to be released from jail in Cromwell's England—and he indeed claims that all differences existing in England are put aside in Barbados. Planters "of the better sort," he informs us, "made a Law amongst themselves, that whosoever nam'd the word *Roundhead* or *Cavalier,* should give to all those that heard him, a Shot [that is, "shoat," a young pig] and a Turky, to be eaten at his house that made the forfeiture; which sometimes was done purposely, that they might enjoy the company of one another."[15]

The gentlemanly conviviality notwithstanding, growing sugar was a difficult business and the tropical environment offered considerable challenges. Ligon returns to the weeding problem in his extended discussion of the sugarcane; he "saw by the growth, as well as by what I had been told, that it was a strong and lusty Plant, and so vigorous, as where it grew, to forbid all Weeds to grow very neer it; so thirstily it suck't the earth for nourishment, to maintain its own health and gallantry." Sugarcane, like most commercial crops, is a fast-growing competitive plant that is highly efficient at turning sunlight and nutrient into tissue and sugar. The biochemistry of this plant or any other was unknown in the seventeenth century, but farmers had certainly understood the importance and effect of fertilization for many millennia. The main issue in maintaining soil fertility is to make available sufficient quantities of the key nutrients nitrogen, phosphorous, and potassium. As Mark Overton has noted, the limiting factor in the early modern period tended to be availability of nitrogen. "Early modern farmers were, of course, ignorant of the existence of nitrogen, but they were nevertheless aware of strategies to maintain fertility which, although they did not realize it, involved the conservation of nitrogen." In early modern Britain, these strategies included fallowing, which allowed nitrogen levels to recover naturally to some extent; crop rotation, particularly of peas and beans, which are able to "fix" nitrogen directly from the air; and grazing animals in pasture during the day (when they ate grass) and folding them in arable fields at night (where they manured the field). Likewise, early modern techniques for controlling weeds and pests were rudimentary and labor-intensive; "in the early modern period," argues Overton, "farmers were comparatively helpless in combating these problems." Those solutions that did exist involved plowing out perennial weeds, weeding out annuals, and picking off caterpillars. These were the long-established techniques of fertilizing, weeding, and pest control that colonists took with them to seventeenth-century Barbados.[16]

Understanding the general importance of a fertile soil was not enough. Temperate and tropical soils are subject to different chemical and physical processes that mean they retain nutrients at different rates and in different proportions. The science behind this was not known, and it took trial and error before planters could work out the most effective fertilization regimes. Likewise, European knowledge about planting wheat, barley, peas, and beans was of limited utility to colonists seeking to identify the specific and correct way to plant the sugarcane rhizome in the soil. This turned out to be no haphazard task. By the early eighteenth century, planters had introduced a precise but back-breaking technique known as "cane-holing." This method, which gave sugar plantations their distinctive gridlike appearance, ensured the cane was correctly aligned and well rooted in the soil, that manure was retained close to the cane, and that the soil was protected from erosion. Before its introduction, however, several different techniques were tried, most involving planting in trenches or furrows in the manner of European cultivation. Ligon described an early attempt at cane-holing that failed and recommended trench-planting instead, but rather than simply contrast the failed and the successful methods, he instead chose to deliver the message in a form reminiscent of a biblical parable. Although understanding the needs of the sugarcane well, the planters

> did not rightly pursue their own knowledge; for their manner was, to dig small holes, at three foot distance, or there about, and put in the Plants endwise, with a little stooping, so that each Plant brought not forth above three or foure sprouts at the most, and they being all fastned to one root, when they grew large, tall, and heavy, and stormes of winde and rain came, (and those raines there, fall with much violence and weight) the rootes were loosened and the Canes lodged, and so became rotten, and unfit for service in making good Sugar. And besides, the roots being far assunder, weedes grew up between, and worse then all weeds, Withs, which are of a stronger grouth then the Canes, and do much mischiefe where they are; for, they winde about them, and pull them down to the ground, as disdaining to see a prouder Plant than themselves.

Ligon's chief fear is the withe, any of several fast-growing vines in the Vitaceae (grapevine) family that spread rapidly, choking other, less vigorous plants. The passage brings to mind both the Parable of the Sower and the Parable of the Wise and Foolish Builders. In the former, Christ tells of a man who sowed seeds. Some fell on stony ground, some among thorns, and some on good ground. In the explanation that follows, "he that received seed into the good ground is he that heareth the word, and understandeth it." Ligon reverses the parable, making it literal once more. The planters had heard how to plant the cane but had not fully understood what they had heard. This caused them to plant the canes badly so that they grew up with weak stems, prone to "stormes of winde and rain." This phrase alludes to the parable about "a foolish man, which built his house upon the sand: And the rain descended, and the floods came, and the winds blew, and beat upon that house; and it fell: and great was the fall of it." Again, Ligon takes the literal meaning of the parable and uses it to imply the foolishness of using the wrong planting technique. Although the battle that ensues is ecological, Ligon personifies the strong, proud, mischief-making withes that compete with the cane to give the impression that these are supernaturally willful weeds, perhaps inspired with some demonic purpose. This is a kind of creation myth of the early sugar planters in which they are granted knowledge, act unwisely, and suffer the consequences. Unlike Eden, however, Barbados is a garden that can successfully be cultivated without fear of expulsion. Ligon provides a long explanation of what he thought to be the correct method of laying the sugarcane rhizomes in the ground before moving on to the next stage, which takes place after about a month, at which point "you shall perceive them to appear, like a land of green Wheat in *England,* that is high enough to hide a Hare." This proverbial measure is still used in England today by farmers deciding when to apply pesticides and herbicides. For Ligon, it signaled a return to weeding:

> But upon the first months growth, those that are carefull, and the best husbands, command their Overseers to search, if any weeds have taken root, and destroy them, or if any of the Plants fail, and supply them; for where the Plants are wanting, weeds will grow; for, the ground is too vertuous to be idle. Or, if any Withs grow in those vacant places, they will spread very far, and do much harm, pulling down all the Canes they can reach to. If this husbandry be not used when the Canes are young, it will be too late to finde a remedy; for, when they are grown to a height, the blades will become rough and sharp in the sides, and so cut the skins of the Negres, as the blood will follow; for their bodies, leggs, and feet, being uncloathed and bare, cannot enter the Canes without smart and losse of blood, which they will not endure. Besides, if the Overseers stay too long, before they repair these void places, by new Plants, they will never be ripe together.

The mythic quality that had characterized the episode about the "foolish" planters here gives way to a passage written in a clear and precise tone. It contains one of the few references to enslaved field laborers anywhere in Ligon's discussion of the sugarcane. Although he by no means suggests that slavery is wrong or that slaves should be freed, this is nonetheless a relatively humane passage in which his primary reason for recommending early weeding and replanting is that leaving it until later will cause injuries among the enslaved people ordered to do the work. The purely agricultural reason, that the canes should ripen at the same time, is a secondary consideration. This is not an antislavery argument in any sense, but it is broadly ameliorative. Importantly, and perhaps ironically, Ligon's consideration for the welfare of the enslaved people under his command emerges from his proto-ecological understanding of the competition between plant species, and its effect on the growth of the crop.[17]

LIGON AND SLAVERY

Ligon's complex and often contradictory attitudes toward slavery and enslaved people are found at numerous locations throughout the *History*, especially in the second half of the book, which is largely structured as a natural history. This is hierarchical, adopting the conventional arrangement of the great chain of being. It starts with the human sphere, beginning with the island's recent history, its geography, and its main economic activities, in particular, food production, before setting out an ethnography of its human inhabitants. It then moves on to "Tame beasts," then "Birds," followed by "lesser Animals and Insects," and finally the island's plant life. It is in this final section that Ligon describes in detail the propagation, management, and processing of sugarcane in which he recommends weeding strategies that prevent injuries among the slaves, but his discussion of the island's enslaved population takes place across the book. Ligon's ethnography is fascinating not just for its portrait of Barbadian society in the mid-seventeenth century, but also because it reveals that his tentative and fluid sense of racial identity is much less clear than his understanding of social structure. Modern readers, viewing the seventeenth century through the lens of later pseudoscientific racial hierarchies that placed Europeans ahead of others, might expect Ligon's account of the human inhabitants to be so ordered, but it is not. A few paragraphs on European "servants," by which he means indentured laborers, is followed by a very long section on "Negroes," a shorter section on "Indians," and a few pages on "Masters." Ligon's account of the inhabitants of Barbados is thus arranged more by his perception of social order than of racial hierarchy, with European indentured servants separated widely from European masters. Ligon's ethnography is also notable for its humanity and sense of justice, at least by early modern standards. There is no question that Ligon saw Africans, Native Americans, and the poor Europeans who indentured themselves as being on the same social footing. Nevertheless, he condemns cruelty while at the same time appearing to recognize that all human beings are equal before God. This emerges through stories that illustrate African and Native American capacity and quietly encourage planters to adopt fair and—relatively—forgiving policies toward enslaved people. Ligon does not condemn slavery per se, nor do we know how he behaved in real life, but in his *History*, he represents enslaved people, whatever their origin, as equal members of the human family.

The first inkling we get of Ligon's attitude toward slavery is his mordant observation that "the slaves and their posterity, being subject to their Masters for ever, are kept and preserv'd with greater care than the servants, who are theirs but for five years, according to the law of the Iland. So that for the time, the servants have the worser lives, for they are put to very hard labour, ill lodging, and their dyet very sleight." This may or may not be true, but there can be little doubt that planters extorted as much work as possible from both indentured servants and chattel slaves. It may appear that this passage reflects Ligon's affinity with his fellow Europeans over the island's enslaved Africans, but in the pages that follow, Ligon recounts several personal conversations with named Africans and Native Americans while never speaking of European indentured servants in anything other than general terms. These conversations are represented in vignettes that generally carry a moral that either asserts the humanity of the story's subject or expresses anxiety or skepticism about slavery in some way. The first concerns Ligon's view of the artistic and intellectual capacity of Africans. Ligon was a musician who took the trouble of bringing his theorbo, a type of large lute, on the hazardous sea voyage from England to Barbados. Intrigued by the African drum music he heard performed on Barbados, which he considered "a pleasure to the most curious eares," he attempted to teach an enslaved person called Macaw to play his theorbo. Macaw attended closely and a few days later, Ligon found Macaw practicing on a different, improvised instrument. In telling the tale, Ligon correlates art and nature, calling upon the reader to see Macaw as a natural genius but deliberately obscuring the boundary between learned and innate behavior. The story takes place in a grove of plantain trees, reminding us of a conventionally allegorical garden scene, while the explicit moral establishes it firmly as a parable. Ligon describes showing Macaw his theorbo in "the Plantine grove," and then later:

> I found this *Negro* (whose office it was to attend there) being the keeper of that grove, sitting on the ground, and before him a piece of large timber, upon which he had laid crosse, sixe Billets, and having a handsaw and a hatchet by him, would cut the billets by little and little, till he had brought them to the tunes, he would fit them to; for the shorter they were, the higher the Notes which he tryed by knocking upon the ends of them with a sticke, which he had in his hand. When I found him at it, I took the stick out of his hand, and tried the sound, finding the sixe billets to have sixe distinct notes, one above another, which put me in a wonder, how he of himselfe, should without teaching doe so much. I then shewed him the difference between flats and sharpes, which he presently apprehended, as between *Fa,* and *Mi* and he would have cut two more billets to those tunes, but I had then

> no time to see it done, and so left him to his own enquiries. I say this much to let you see that some of these people are capable of learning Arts.

Karen Ordahl Kupperman has pointed out that Macaw would no doubt have been familiar with the lute, varieties of which were traditional in West Africa, and that here he was almost certainly constructing another traditional African instrument, probably "either a kalimba or a marimba." Kupperman's terminology is a little inexact—the kalimba is a mid-twentieth-century variant of the southern African mbira, for example—but her general point that Macaw would have been familiar with a wide range of African stringed and percussion instruments remains accurate. Ligon did not know this, nor does he appear to have taken the trouble to find out. In this respect he underestimates the complexity of African society, while his assumption that Macaw is simply attempting to imitate Western music reflects the widespread ignorance of African cultures among English settlers in the Caribbean that, in this point at least, he shared. Despite this, however, Ligon's message is more positive. Rather than simply dismissing Macaw's creativity as no more than naïvely imitative, he instead tells the tale to assert the explicit point "that some of these people are capable of learning Arts." Abolitionist writers would later use precisely the same argument in their campaign to dispel racist propaganda that Africans were incapable of anything more than physical labor. Thus, for example, in 1784 the abolitionist James Ramsay, who had read Ligon, would dedicate no fewer than sixty-five pages of his important *Essay on the Treatment and Conversion of African Slaves in the British Sugar Colonies* to the topic "Natural Capacity of African Slaves Vindicated." Throughout the period of abolition and emancipation, white abolitionists would support and promote Black antislavery writers and campaigners, from Ignatius Sancho and Phillis Wheatley to Mary Prince and Frederick Douglass, in part, at least, to reinforce the point "that some of these people are capable of learning Arts." Ligon's theorbo parable, therefore, directly anticipates one of the most important tactics of abolitionist rhetoric.[18]

Ligon locates this vignette in a plantain grove, leading Pluymers to observe that "for Ligon, it was not a purely functional landscape but also a site for leisure and exploration." The plantain grove was a semi-natural setting in which the boundaries between art (as agriculture) and nature are blurred. The plantain (a variety of banana; *Musa* spp.) is an exotic plant for seventeenth-century English readers, but one that is nonetheless cultivated rather than natural, at least in the Caribbean. It is frequently depicted in literature of this period as characteristic of Africa and African society—although its origins are in fact originally Southeast Asian. As an enslaved person, Macaw is himself managed by the overseers as a part of the plantation, but Ligon's plantain grove setting also emphasizes that he is no savage living in a state of nature, but rather the inhabitant of an agricultural society, albeit an exotic one for English readers. This too would become a recurring image in abolitionist literature of the eighteenth century, in which Africans would be frequently represented as peaceable inhabitants of managed plantain groves. For example, Thomas Day, in the preface to the important poem *The Dying Negro* (1773–75), would complain that "the Negro is dragged from his cottage, and his plantane shade" by American slave traders. A few years later, William Roscoe, in his abolitionist poem *The Wrongs of Africa* (1787–88), would describe the "silent groves, / Where palms, and plantains, intermix'd their shade." Here he located his hero, the African warrior and farmer Matomba, whose plantain grove was "a chosen spot, / That own'd his constant culture." In both cases, as in Ligon's parable, the plantain grove is a marker both of African agriculture and of wider African culture.[19]

Ligon's interest in this culture is augmented by his thinking about the law and the status of enslaved people. This is drawn out in his story about Sambo, who asked Ligon how the compass worked. The explanation "put him in the greatest admiration that ever I saw a man, and so quite gave over his questions, and desired me, that he might be made a Christian; for, he thought to be a Christian, was to be endued with all those knowledges he wanted." Ligon attempted to comply with Sambo's request:

> I promised to do my best endeavour; and when I came home, spoke to the Master of the Plantation, and told him, that poor *Sambo* desired much to be a Christian. But his answer was, That the people of that Iland were governed by the Lawes of *England,* and by those Lawes, we could not make a Christian a Slave. I told him, my request was far different from that, for I desired him to make a Slave a Christian. His answer was, That it was true, there was a great difference in that: But, being once a Christian, he could no more account him a Slave, and so lose the hold they had of them as Slaves, by making them Christians; and by that means should open such a gap, as all the Planters in the Iland would curse him. So I was struck mute, and poor *Sambo* kept out of the Church; as ingenious, as honest, and as good a natur'd poor soul, as ever wore black, or eat green.

The reference to eating green is mysterious, but in general this vignette offers another example of the way in which Ligon wants his readers to see him as reasonable and compassionate. He both identifies the moral hypocrisy at the heart of Christian slaveholding and promotes the intellectual capacity and moral honesty of an enslaved African. The passage would not have been out of place in abolitionist tracts published more than a century later.[20]

Another of Ligon's vignettes, this one in which enslaved people compete to catch a Muscovy duck (*Cairina moschata*), is akin to fable—an extended metaphor with an animal at its center—but presented as a spectacle in which enslaved people are ordered to perform for the entertainment of a few onlookers. Ligon begins by asserting that "excellent Swimmers and Divers they are, both men and women" but, almost immediately, the scientific purpose of the passage is, literally, submerged. "Collonell *Drax*," recounts Ligon, "would sometimes, to shew me sport, upon that day in the afternoon, send for one of the *Muscovia* Ducks, and have her put into his largest Pond, and calling for some of his best swimming Negres, commanded them to swim and take this Duck." To make things more challenging, Drax prohibited diving to take the duck from underneath. Ligon was content to enjoy this sport. Indeed, he notes, "in this chase, there was much of pleasure, to see the various swimmings of the *Negroes;* some the ordinarie wayes, upon their bellies, some on their backs, some by striking out their right legge and left arme, and then turning on the other side, and changing both their legge and arme, which is a stronger and swifter way of swimming, then any of the others." The description of the various strokes adopted by the African swimmers is also perhaps an attempt to recover the scientific rationale for the vignette—the unfamiliar stroke he describes is now called the front crawl, which was not introduced into British swimming until the 1840s. Nevertheless, Ligon is ultimately more interested in a moral story about blame, forgiveness, and fairness than he is in swimming strokes:

> While we were seeing this sport, and observing the diversities, of their swimmings, a *Negro* maid, who was not there at the beginning of the sport; and therefore heard nothing of the forbidding them to dive, put off her peticoate behind a bush, that was at one end of the Pond, and closely sunk down into the water, and at one diving got to the Duck, pul'd her under water, & went back againe the same way she came to the bush, all at one dive. We all thought the Duck had div'd: and expected her appearance above water, but nothing could be seen, till the

> subtilty was discovered, by a Christian that saw her go in, and so the duck was taken from her. But the trick being so finely and so closely done, I begg'd that the Duck might be given her againe, which was granted, and the young girle much pleased.

This story of hunting underwater echoes the description of seabirds, dolphins, and flying fish that Ligon had offered earlier in the book. Modern readers are likely to find this passage disturbing; a reminder that enslaved people could be commanded to play, hunt, swim, or work at their masters' pleasure, or simply for their entertainment. Even Ligon's advocacy on behalf of the young girl reminds us that justice, such as it was, was at the whim of masters and overseers. Contemporary readers, however, may have been "much pleased" that innocence and ability were rewarded. More to the point, Ligon might have intended that his reader see him in a positive light as an advocate for a young girl who was unfairly treated—and perhaps remember while reading that the author himself was writing this passage in jail, after unfairly losing his own property at the hands of others. Either way, this is an animal story with a moral—in this case that ignorance sometimes is a defense—and as such can be described as a fable. It reminds the reader of the abilities of Africans, while hinting that they too should benefit from fair treatment. At the same time, it sits within a natural history. The animal story says something useful about the wildlife of Barbados as well as about its human inhabitants. Finally, because it brings so strongly to mind Ligon's earlier account of the fish and hunting birds of the Atlantic, it asks the reader to think of the plantation as a whole as a kind of ecosystem akin to, as he had put it earlier, "a well governed Common-wealth."[21]

The Macaw, Sambo, and Muscovy duck episodes, taken together, present one side of Ligon's approach to the enslaved people of Barbados. His overall position was, however, far more complex and contradictory and contains elements of the emerging racial discourses that would later be used to justify slavery. In particular, Ligon repeatedly contrasts, and sometimes compares, Africans with domesticated animals. Despite his insistence on "their aptnesse to learne Arts," for example, Ligon appears unashamed when noting that, when purchasing enslaved people, planters "choose them as they do Horses in a Market; the strongest, youthfullest, and most beautifull, yield the greatest prices." He is similarly blunt at the end of the passage when asserting that every enslaved person in the market "knowes his better, and gives him the precedence, as Cowes do one another, in passing through a narrow gate; for, the most of them are as neer beasts as may be, setting their souls aside." The market passage is somewhat confused, reflecting Ligon's ambivalent and anxious discussion of slavery throughout the book, but also reminding us that seventeenth-century racial thinking was largely tentative and unformed. On the one hand, he offers these two animal similes to describe slaves in the market. On the other hand, these crude representations are undercut by the human reality of the situation. Unlike animals, the men wish to take wives and when they are prevented from doing so, "the men who are unmarried will come to their Masters, and complain, that they cannot live without Wives, and desire him, they may have Wives. And he tells them, that the next ship that comes, he will buy them Wives, which satisfies them for the present." Despite comparing the men with animals, Ligon explicitly denies that they are, with the final line asserting that they have human souls. Whether as Africans, as slaves, or as both, Ligon conceives of the men as *like* animals, but not *as* animals themselves.[22]

Similar doublethink would be exhibited in the writings of planters and slave owners for centuries to come, but few would exhibit Ligon's ambivalence. In an extraordinary passage a little later in the book, for example, he likens an enslaved African woman working in a field to a warhorse bearing on her back no less a knight than St. George, slayer of dragons and the patron saint of England. "Some women," he relates, "whose Pickaninnies are three yeers old, will, as they worke at weeding, which is a stooping worke, suffer the hee Pickaninnie, to sit astride upon their backs, like St. *George* a horse back; and there spurre his mother with his heeles, and sings and crowes on her backe, clapping his hands, as if he meant to flye; which the mother is so pleas'd with, as shee continues her painfull stooping posture, longer then she would doe, rather than discompose her Joviall Pickaninnie of his pleasure, so glad she is to see him merry." The passage is evidently intended to be a jovial and merry vignette, although the child's joy comes at the cost of the mother's pain. Her representation as a beast of burden is consistent with similar representations throughout centuries of plantation literature, but the depiction of her son is unusual. He is compared with one of Christianity's most venerated saints, but this gives way to a description of him as, first, a songbird that sings and next a cock, or crow, with a more discordant call. His attempt to fly is comically unsuccessful but, given that Ligon is a keen observer of birds as well as people, it is hard to read this as malicious. It is, indeed, one of the earliest of many comparisons between birds and enslaved people to appear in English colonial literature. Most would draw on birds as symbols of freedom, and caged birds as metaphors for enslavement. This passage does not quite anticipate the tradition that was yet to develop, but it alerts us to Ligon's complex and divided relationship both with the animals and with the enslaved people around him.[23]

A few pages on, Ligon continues his description of African women in another extraordinary passage that has attracted the interest of several critics including Susan Dwyer Amussen, who notes that "the strangeness of Barbados permitted Ligon to write about enslaved women in a way that would be unthinkable in relation to English women of any class." He begins by observing that African men are "very well timber'd that is, broad between the shoulders, full breasted, well filleted, and clean leg'd, and may hold good with *Albert Durers* rules, who allowes *twice the length of the head,* to the breadth of the shoulders; and twice the *length of the face,* to the breadth of the hipps." If African men meet with his aesthetic approval, the same cannot be said of African women who, he says, "are faulty; for I have seen very few of them, whose hipps have been broader than their shoulders, unlesse they have been very fat. The young Maides have ordinarily very large breasts, which stand strutting out so hard and firm, as no leaping, jumping, or stirring, with cause them to shake any more, then the brawnes of their armes. But when they come to be old, and have had five or six Children, their breasts hang down below their navells, so that when they stoop at their common work of weeding, they hang down almost to the ground, that at a distance, you would think they had six legs." Amussen notes that "Ligon moves to imagining a new beast. African women were another of Barbados's bizarre animals." This analysis is rhetorically satisfying, but not quite convincing. The description certainly does occur in a natural history, but Ligon does not represent the island's other animals as particularly bizarre and it is not at all clear that he thinks African women are anything other than human. Jennifer Morgan has more persuasively argued that the women portrayed in this passage are human, but monstrous. "In this context," she suggests, "black women's monstrous bodies symbolized their sole utility—the ability to produce both crops and other laborers." She notes that the passage contrasts strongly with the representation of the beautiful Black women he encountered on Cape Verde, but she sees this as part of a deliberate narrative continuum. "Taking the female body as a symbol of the deceptive beauty and ultimate savagery of blackness," she argues, "Ligon allowed his readers to dally with him among beautiful black women, only to seductively disclose their monstrosity over the course of the narrative." This is a convincing reading, particularly since Ligon cites Albrecht Dürer's 1528 *Four*

Books on Human Proportion at the outset of both discussions, arguably bringing them together into a single unified narrative. Morgan does not draw attention to this, however, instead making the broader point that "Ligon's narrative is a microcosm of a much larger ideological maneuver that juxtaposed the familiar with the unfamiliar—the beautiful woman who is also the monstrous laboring beast." She concludes that "in the discourse used to justify the slave trade, Ligon's beautiful negro woman was as important as her 'six-legged' counterpart. Both imaginary women marked a gendered and, as Kim Hall has argued, a stabilized whiteness on which European colonial expansionism depended." We might also note that in Ligon's account, the transformation from beauty to monstrosity occurs as part of the translocation from African island to Caribbean island. The natural beauty that he describes in the women of Santiago becomes unnatural, monstrous, and deformed in Barbados. Nevertheless, Ligon was by no means the first to describe African women in this way. William Towerson, for example, describes African women's breasts in exactly these terms in his mid-sixteenth-century account of Africa, which was reproduced in Richard Hakluyt's *Principall Navigations, Voiages, and Discoveries of the English Nation,* published when Ligon was a young man. "Divers of the women have such exceeding long breasts," says Towerson, "that some Long women of them wil lay the same upon the ground and lie downe by them." Their breasts, he had said on the previous page, "in the most part be very foule and long, hanging downe low like the udder of a goate." Ligon was almost certainly familiar with images such as these that represented African women as akin to domesticated animals.[24]

While these representations apparently articulate a submerged anxiety about the effect of slavery, they are far from being statements of antislavery. There can be no doubt that Ligon wished success to British colonists in Barbados, or he would not have gone so far out of his way to provide them with such useful information about the construction and maintenance of a sugar plantation. It is also clear that his representations of Africans, women in particular, "juxtaposed the familiar with the unfamiliar" in ways that emphasized difference, even monstrous difference. Sometimes, however, as with his depictions of Macaw and Sambo, they emphasized common humanity, although he is generally readier to recognize common humanity in men than in women. It is also difficult to argue that Ligon explicitly attempts to justify the slave trade or that he was an enthusiast for plantation discipline. The evidence of his writing is that he was not wantonly cruel. Even when he compares enslaved people to animals, he does not imply that he thought they should be subjected to cruel treatment, any more than he thought animals themselves should be abused. Indeed, in several places in the *History* he explicitly advocates humane treatment of animals, as for example when he recommends pigs have larger sties, when he sympathizes with the turtle he is about to kill and eat ("when he sees you come with a knife in your hand to kill him, he vapours out the grievousest sighes, that ever you heard any creature make, and sheds as large tears as a Stag"), or when he decries horse racing, a sport that for him is "forcing poor beasts beyond their power, who were given us for our moderate use." The key word is "moderate." Throughout the *History*, Ligon attempts to walk a middle path that both recognizes the essential humanity of the enslaved people he encounters, even when he perceives them as unnatural or deformed, and supports the colonial ambitions of the British planters in Barbados. Ultimately, however, such an attempt would always be doomed to fail. The truth was that even planters who thought themselves moderate in the use of coercive violence were still, quite literally, working their slaves to death.[25]

Whether one characterizes Ligon's conflicted views as moderation, ambivalence, doubt, or even calculated policy, they opened a space for later readers to ponder the morality of slavery. The *History* remained an authority for many years. It was consulted and summarized by James Ramsay as late as the 1780s in manuscript notes and, on the other side of the debate, cited by the proslavery lobby group known as the West India Interest in their evidence to the Privy Council in 1788. It remained visible through numerous citations in both natural history and abolitionist literature, including such central texts as Hans Sloane's *Voyage to Jamaica* and Anthony Benezet's Some *Historical Account of Guinea*—even if, as early as 1708, John Oldmixon (1673–1742) found him "old." Nevertheless, Richard Steele was still reading him in 1711 and it was through Steele that Ligon found an enduring legacy in popular literature. Ultimately, Ligon's most important contribution to public discourse was the story of Yarico's enslavement at the hands of an unfeeling and rapacious English merchant who would become known to later generations as Thomas Inkle. Steele did not merely give the young man a name. He also, albeit with generous helpings of irony, recast the story as a problem with nature in both its human and its environmental senses. Steele's Inkle "was the third Son of an eminent Citizen, who had taken particular Care to instill into his Mind an early Love of Gain, by making him a perfect Master of Numbers, and consequently giving him a quick View of Loss and Advantage, and preventing the natural Impulses of his Passions, by Prepossession towards his Interests." If greed allows him to master his natural passions, it is nature that undoes his self-control. He encounters Yarico in "a remote and pathless Part of the Wood," from where she "conveyed him to a Cave, where she gave him a Delicious Repast of Fruits, and led him to a Stream to slake his Thirst." These natural refreshments are augmented by Yarico's use of the unrefined products of nature to adorn herself and Inkle's cave. Living in a state of nature, she "every day came to him in a different Dress, of the most beautiful Shells, Bugles, and Bredes. She likewise brought him a great many Spoils, which her other Lovers had presented to her; so that his Cave was richly adorned with all the spotted Skins of Beasts, and most Party-coloured Feathers of Fowls, which that World afforded." In case the pelts and plumage of dead animals were not sufficiently alluring, she introduces Inkle to the charms of running water and living birds by taking him "in the Dusk of the Evening, or by the favour of Moon-light, to unfrequented Groves, and Solitudes, and show[ing] him where to lye down in Safety, and sleep amidst the Falls of Waters, and Melody of Nightingales." It is only after leaving this Edenic wonderland and returning to civilization that Inkle recovers his love of gain and sells Yarico into slavery. Inspired by Ligon's story, but embellishing it considerably, Steele's tale of Inkle and Yarico argues that slavery is incompatible with life in a state of nature and firmly posits slave trading as a corruption of human nature arising from a culture that commodifies the living world, both human and nonhuman.[26]

Already a fully formed antislavery parable in Ligon's *History*, the tale of Inkle and Yarico went on in its various retellings to become one of the most important reminders for eighteenth-century readers that slavery was a nasty and unnatural business, corrupting slave traders just as much as destroying the lives of those who were enslaved. Thanks to Yarico, Ligon's contribution to colonial discourse was probably more substantial than is often recognized, setting many of the terms by which both Caribbean nature and Caribbean slavery would be understood and discussed in the eighteenth century. The central point is that his discussions of enslaved Africans sit within the ethnographic section of a natural history. Later readers like Steele, interested in the flora, fauna, and agriculture of Barbados, were also exposed to Ligon's views on enslavement. While often contradictory, and certainly not explicitly opposed to slavery, Ligon's observations certainly supported arguments about the natural ability of Africans that were promoted by later antislavery campaigners. Ligon's broadly ameliorationist ideas are also matched by arguments that could be described as proto-ecological. His worldview was one in which reciprocal social relations were mirrored by reciprocal biological relationships, in which there was a place for everyone and everything and in which all things played an important part. While this might sound like modern green utopianism, it was in fact not far removed from traditional Christian ideas about society and the state. Ligon arguably did little more than to add his observations as a natural historian to an existing organic conception of society, tempered by what seems to have been his own genial personality. And yet, taken together, these approaches added up to a worldview that was holistic, egalitarian, and, in a rudimentary way at least, skeptical about slavery. He had written a book that could both undermine a reader's confidence in slavery as an institution and augment a reader's understanding of the interrelatedness of the natural world. For these reasons it remained—and remains—one of the foundation stones of British Caribbean writing.

CHAPTER 2

"A very perverse Generation of People"

NATURAL HISTORY IN THE SERVICE OF THE PLANTERS

In 1727, the brief but rhapsodic *Some Modern Observations upon Jamaica: As to Its Natural History, Improvement in Trade, Manner of Living, &c* appeared as part of a collection of poems, letters, and other documents associated with Philip Wharton (1698–1731), the Jacobite first Duke of Wharton. The anonymous author enthusiastically promoted colonization of Jamaica, starting from a solid endorsement of natural theology. Witnessing the island from the sea, this author suggests that "the *Atheist* might be tempted to exult, and his foolish Heart imagine the Finger of God had not been here." However, once on shore the experience of witnessing extraordinary tropical diversity will soon change his mind: "the impious Thought will soon be confounded by wonders of the animal and vegetable Worlds, to crush his Soul with ten thousand invincible Demonstrations." Having established God's soul-crushing bounty, the author considers ways to profit from it. He notes that there are "very rich Mines" in Jamaica that are ripe for development. "Such an Enterprize might be undertaken upon very rational Grounds," he argues, "and methinks the cheap Labour of Slaves, and almost every where Woods that may be had for nothing, ought greatly to encourage it." Whether trees, mines, or slaves, Jamaica offers resources for the taking. That these slaves are human beings goes unmentioned. In this author's view, enslaved people are simply another natural resource to be exploited.[1]

The pamphlet may be brief and coarse, but the attitudes it displays were proliferating in the first half of the eighteenth century. As the ambivalence that characterized the writing of early colonists such as Richard Ligon was swept away, both natural history and plantation management literature increasingly worked to naturalize slavery to the Caribbean, consolidating and extending the view that the enslavement of Africans was the natural response to the opportunities presented by a bountiful Caribbean environment. This perspective become dominant by the mid-eighteenth century, represented in written forms of heightening sophistication, although ever more distanced from the estates that they paradoxically portray with increasing detail and scientific awareness. By 1750, readers interested in the fauna and flora of the Caribbean were also exposed to a set of attitudes about enslaved people that tended to justify slavery as the normal and natural social arrangement of the islands—such that we can speak of natural history in this period and region as having been written in the service of the plantocracy. Two substantial natural histories produced between 1707 and 1750 demonstrate this. The first, Hans Sloane's *Voyage to Jamaica,* was considered a major work in its time and continues to be the subject of extensive scholarly interest. The second is Griffith Hughes's *Natural History of Barbados,* published with great fanfare in London in 1750, but quickly derided for its scientific shortcomings. Both, in their various ways and with differing degrees of success and authority, contributed to the eighteenth-century discourse of the naturalized plantation and both provided raw materials from which later abolitionists were able to fashion the narrative that slavery was not naturalized but unnatural.

HANS SLOANE'S VOYAGE TO JAMAICA

By the early eighteenth century, Jamaica was overtaking Barbados as the most valuable British colony in the Caribbean. As well as offering plentiful opportunities for planters, its size also meant that it had a wider range of habitats and greater biodiversity than the smaller islands. It thus became an important object of study for scientists and was the site of one of the most important natural histories of the eighteenth century: Hans Sloane's *A Voyage to the Islands Madera, Barbados, Nieves, S. Christophers and Jamaica,* published in two large volumes in 1707 and 1725. Sloane remains relatively well known today, memorialized in London's fashionable Hans Town and Sloane Square and celebrated as the man who reinvigorated the Royal Society after Isaac Newton's long presidency, who introduced milk chocolate to Great Britain, and who bequeathed the collections that would ultimately become the British Museum, British Library, and the Natural History Museum. Born in 1660 into the Irish Protestant Ascendancy in Killyleagh, County Down, he lived an astonishingly productive life as a physician, scientist, and collector until his death in London in 1753 at the age of ninety-two. In the late 1680s, as a young man, Sloane spent a little over a year in Jamaica, and in 1695 he married Elizabeth Langley Rose (1662–1724), through whom he inherited substantial plantations in Jamaica and thus himself became a slaveholder. Together, these experiences would lead to and inflect the two volumes of his great work, which present more than two thousand pages of detailed and precise description of Caribbean flora, fauna, geography, climate, and society and which quickly became one of the primary sources of information not only for Jamaican flora and fauna but also for the island's plantation agriculture and its predominantly enslaved population.[2]

Sloane fundamentally altered both the practice and the style of Caribbean natural history with his exhaustive and detailed account of Jamaica's flora and fauna. The *Voyage* was significant in being the first colonial natural history to apply the taxonomic system that had been outlined by the naturalist John Ray (1627–1705) in his *Methodus plantarum nova,* or *New Method of Plants* (1682), and further developed in his *Historia plantarum*, or *History of Plants,* the first volume of which had appeared in 1686, the year before Sloane set sail for Jamaica. Ray introduced a taxonomy based on close observation and comparison of the physical features of plants, such as the number of petals on the flower. He was also the first to speak of organisms as "species" in the truly scientific sense, famously noting that the seed of one species never gives rise to an individual of another species. Ray's method was highly influential, not least on Linnaeus, whose system of binomial nomenclature was ultimately adopted by the international scientific community, but whose comparative method owed much to Ray. Sloane explicitly cites his debt to Ray, to whom he showed an early draft of the *Voyage,* and his careful use of Ray's taxonomy locates his *Voyage* as the earliest systematic natural history of the Caribbean. Although the book attempts to be impartial, in that it balances differing points of view where they exist, we should nevertheless be cautious of judging it—or indeed any book of this period—by modern standards of scientific objectivity which, in the view of historians of objectivity Lorraine Daston and Peter Galison, is "knowledge that bears no trace of the knower." Sloane is happy to do experiments on himself or to record, anecdotally, his own symptoms or the circumstances by which he arrived at a discovery. For example, many of the opening pages describe his seasickness on the voyage out. As a doctor and a scientist, he turns this unpromising start to account, beginning with a dissertation on the causes, symptoms, and cures for motion sickness. While this is an important medical question for a maritime nation, the blend of autobiography and scientific inquiry reminds us that his scientific method uses different criteria for objectivity than would be current today. Nevertheless, while Sloane is present as a personality throughout the *Voyage,* he in general tries to describe colonial slavery without either defending it or urging its amelioration, even though the attempt is not entirely successful. Consciously or otherwise, the attempted impartiality of Sloane's *Voyage* tends to represent slavery as simply a fact of the Caribbean environment that has few, if any, moral or ethical implications. Sloane's ostensibly scientific language of impartial observation, comparison, and description has the effect, therefore, of representing slavery as a natural, or at least as a naturalized, phenomenon in the Caribbean.[3]

Structurally, Sloane's *Voyage* resembles those of earlier visitors to the Caribbean, beginning as a voyage narrative before evolving into a more systematic natural history. Like Richard Ligon, Sloane passed time on the voyage by watching seabirds hunting for fish. He describes "sea swallows," citing both Ligon and the ornithologist Francis Willughby (1635–72), but while Willughby's sea swallows are almost certainly the common tern (*Sterna hirundo*), which Sloane must have known well, Sloane names the species he observes as *Hirundo Marina major* to distinguish it from the common tern, then called *Hirundo Marina,* a literal Latin translation of "sea swallow." Sloane's description of the bird ("dark grey colour on the Back, and white below") is more consistent with the brown noddy (*Anous stolidus*) than a tern but, identification aside, where he initially differs from Ligon is in his description of the bird's behavior. Ligon had viewed the bird holistically, considering its interrelationships with both its biotic environment of fish and its abiotic environments of water, land, and air. Sloane, by contrast, takes a more reductionist approach. He does describe the bird flying low across the sea surface "being very intent on its Prey, and getting what Fish it could spy there," but he does not at first consider the response of the fish or the involvement of dolphinfish as Ligon had done. Instead, there is a description of the bird followed by a discussion of the existing literature. Fifteen pages further on, however, his discussion is clearly influenced by Ligon, whom he cites twice in connection with dolphins, there being "as much pleasure in seeing them pursue the Flying-Fish, as in Hunting or Hawking." This nod to Ligon's method is not, however, often reprised.[4]

Sloane's method is generally more consistent than Ligon's, but it may not always have been more precise. Sloane tells us in his preface that while in Jamaica he chose the places that interested him, rather than systematically identifying different habitats, and that he estimated everything quite literally by rule of thumb. "Upon my Arrival in *Jamaica,*" he reports, "I took what pains I could at leisure-Hours from the Business of my Profession, to search the several Places I could think afforded Natural Productions, and immediately described them in a Journal, measuring their several Parts by my Thumb, which, with a little allowance, I reckoned an Inch. I thought it needless to be more exact, because the Leaves of Vegetables of the same sorts, Wings of Birds, &c. do vary more from one another, than that does from the exact measure of an Inch." Sloane's decision not to attempt to measure to impossible degrees of precision is certainly justifiable, but his choice of his own thumb as a rule is less defensible since the key elements of scientific observation are standardization and the ability to reproduce one's findings exactly. Sloane's thumb can never be reproduced. Likewise, his habit of simply choosing likely spots to observe vegetation is hardly scientific by today's standards, but in a country new to him almost every corner yielded new discoveries. These Sloane recorded in several ways. "After I had gather'd and describ'd the Plants," he tells us, "I dried as fair Samples of them as I could, to bring over with me. When I met with Fruits that could not be dried or kept, I employ'd the Reverend Mr. *Moore,* one of the best Designers I could meet with there, to take the Figures of them." Sloane also began the collection for which he would become famous. "When I return'd in England," he says, "I brought with me about 800 Plants, most whereof were New, with the Designs beforementioned." Many of these samples and illustrations remain intact at the Natural History Museum in London, reflecting Sloane's interests not merely as a private collector but also as a public benefactor. The book as a whole, says Raymond Stearns, was "a pioneer effort in the writing of the natural history of the New World." He notes that "Sloane's treatments of the fauna were inferior to and less complete than his accounts of the flora," but concludes that "in the absence, however, of any other illustrated natural history of Jamaica, it was a masterly accomplishment." James Delbourgo goes further, calling the book "one of the age's most spectacular works of colonial science."[5]

Collection was not Sloane's only method. He also sought out "the best Informations I could get from Books, and the Inhabitants, either *Europeans, Indians,* or *Blacks.*" He notes that enslaved people "took care to preserve and propagate such Vegetables as grew in their own Countries" and that "I made search after these, and what I found, is related in this History." This is the first indication that Sloane was in contact with enslaved people on the island and suggests that much of his information was derived from African and Native American traditional knowledge. The species he observed, many of which he was the first person to describe, he ordered according to the best taxonomic system of the day. Taken together, these techniques made Sloane's method extremely effective and it became a model for later naturalist-explorers. Its reductionist approach, however, separating off different species and describing them in isolation rather than considering their interactions, meant that his biological understanding, like that of much of the age, was ultimately rather static. Sloane's book describes the wildlife of Jamaica, but it makes little attempt to interpret it.

Sloane's underlying philosophy becomes apparent in a passage attacking the idea of the transmutation of species over time, a proto-evolutionary attempt to explain diversity. His real purpose was probably to justify the importance of his own book rather than to challenge any serious attempt made by others to explain the world, but he nevertheless speaks out plainly, declaring that species "in probability, have been ever since the Creation, and will remain to the End of the World, in the same Condition we now find them." Asserting that natural history affords the naturalist "great Matter of Admiring the Power, Wisdom and Providence of Almighty God, in Creating, and Preserving the things he has created," Sloane concludes that "there appears so much Contrivance, in the variety of Beings, preserv'd from the beginning of the World, that the more any Man searches, the more he will admire; And conclude them, very ignorant in the History of Nature, who say, they were the Productions of Chance." The claim that life in all its diversity had arisen by chance was not frequently made in this period, but it did have a long pedigree, reaching back to ancient Greek atomists such as Democritus (c. 460—c. 370 BCE) and Epicurus (341–270 BCE), who believed the world had coalesced from the collision of tiny particles and from there had increased in complexity. The seventeenth century saw renewed interest in Epicurean ideas, in part stimulated by René Descartes (1596–1650) and those who were inspired by him. By the time Sloane was writing, Deism, the belief that God had set the world in motion and thereafter left it to run according to laws of nature, was gaining a following. Often expressed in somewhat rarefied, intellectual circles, Deism and skepticism were increasingly being popularized in texts such as John Toland's (1670–1722) *Christianity Not Mysterious* (1696), which sought to dispel belief in biblical miracles. At the same time, natural philosophers such as Robert Hooke (1635–1703) were starting to question whether the world had indeed remained unchanged since the creation. Significantly, Hooke asserted that nature was subject to change in a lecture on Caribbean volcanoes published just two years before Sloane's *Voyage*. Sloane clearly rejects such controversies, aligning himself with orthodox Anglican belief and those like Ray, who used observation of natural history as evidence of God's existence and power; an approach then called "physico-theology" but today better known as "natural theology."[6]

Natural theology leads Sloane into ways of thinking that might not have been so significant were he simply cataloguing the species he found rather than taking a view on their origin and purpose in a wider plan. The implication is that everything that is observed must be in the correct place, at the correct time, and in the correct form, because that is where God had placed it—a philosophy powerfully expressed by Alexander Pope (1688–1744) in his *Essay on Man* in the observation that "All Nature is but Art, unknown to thee; / All chance, direction, which thou canst not see." Pope's natural theology famously concludes with the religiously conservative assertion "One truth is clear, WHATEVER IS, IS RIGHT." Although neatly put, Pope oversimplifies a complex moral, philosophical, and theological debate. Nevertheless, he sums up a point of view that was common not only to natural but also to political histories. This view has important implications for the study of slavery. For natural historians, the idea that everything was where God wanted it to be tended to discourage investigation into how things came to be where they were and what the relationships between those things might be. This view was both hierarchical, in that it placed God at the top of his creation, and reductionist, in that it posited that God had created various discrete species and placed them where they belonged and where they stayed. As with species, so with people. The same worldview justified the belief that God had appointed people to a particular social level and could be used to justify the idea that because slavery existed it must therefore be part of God's plan. In this view slavery and the slave trade did not merely serve the local economic interests of the British plantocracy, they were divinely ordained. Slavery might be unfortunate: something to be regretted and its victims pitied, but it was not something that could be changed any more than one species could mutate over time into another. In this worldview, enslaved people, "in probability, have been ever since the Creation, and will remain to the End of the World, in the same Condition we now find them."[7]

This approach was contested by some at the time. It was not difficult to assert that, unlike plants and animals, people had free will and could deviate from God's plan. Slaveholding, in this view, was a wicked perversion of God's intentions, practiced by sinful men. The seventeenth-century pamphleteer Thomas Tryon saw slavery as unnatural because it directly contradicted human nature as God had created it: "Whereas [man] was made a sociable Creature," he argued, "intended for the well-ordering of the inferior Beings, and the help and comfort of those of his own Species by mutural Acts of Benovelence, Courtesie, and Charity: he is now become a *Tyrant,* a *Plague,* a *professed Enemy, Hunter, Betrayer, Destroyer* and *Devourer* of all the Inhabitants of *Earth, Air* and *Water.*" Quakers in Barbados and Pennsylvania had also asserted the sinfulness of slavery by the close of the seventeenth century. It is hard to believe that Sloane would never have encountered the idea that slavery is sinful, but this is not how he describes it in his *Voyage.* Instead, he attempts to describe slavery and enslaved people impartially, neither sanctioning nor censuring the system that he found in Jamaica. By so doing, he implicitly presents slavery as part of the natural order, and enslaved people as figures in a static and divinely sanctioned landscape. He thus represents slavery as a discrete object in the social landscape rather than as part of a global network of social, political, and economic relationships, just as he sees plants as discrete objects in the natural landscape rather than as connected parts of an ecosystem. His discussion merely describes enslaved people as they are, and this, while appearing to have scientific neutrality, is in fact anything but neutral in that it implies that slavery is an immutable natural phenomenon put in place by a wise creator.[8]

This was Sloane's position as he set out for Jamaica, battling seasickness and watching birds along the way. He called first at Barbados, noting that the deforestation that Richard Ligon had witnessed was now complete, leading to a more general environmental degradation of the island. "They at Barbados want Wood very much," he wrote, observing that Barbados "has had so great a fruitfulness, though it be fallen off from what it was, through the great labouring and perpetual working of it out, so that they are now forc'd to dung extremely what before was of it self too Rank." The initial burst of high productivity, caused by releasing the nutrients locked up in the rainforest into the soil, had lasted less than fifty years. The sugar-slave system had destroyed an irreplaceable ecosystem within just a few decades, creating what we might call a "dung crisis" on the island. Although Sloane notes this damage, he does not see it as a loss. In the introduction, he laments, " 'Tis a very strange thing to see in how short time a Plantation formerly clear'd of Trees and Shrubs, will grow foul, which comes from two causes; the one the not stubbing up of the

Roots, whence arise young Sprouts, and the other the Fertility of the Soil." For Sloane, the rainforest in its fertile natural state is foul; plantation, even if it requires extreme manuring, is the more desirable land use. But we should not assume from this that Sloane was blind to environmental damage and pollution. In his description of the Jamaican "fresh-water mullet" (which he identifies as *Mugil cephalus,* today known as the striped mullet or flathead grey mullet), he notes the fish are "very good and delicious Food, being extreamly fat and savory, which may come from the Rivers not being here foul'd with excrementitious Matters so much as those of *Europe.*"[9]

Sloane's introduction is in many ways the most interesting part of his book for the historian of slavery. In these pages he is less the specialized botanist and more the interdisciplinary environmental scientist, concerned with the geography, geology, climate, and agriculture of the island as well as its social composition and political history. As a doctor, he is also very interested in tropical diseases. Many of the cures he recommends derive from plants, making his botany a matter of medical importance as well as scientific curiosity, and indeed many of his descriptions of plants incorporate discussion of plants' medicinal "virtues" in the manner of traditional herbals and pharmacopoeias. His introduction also contains a substantial ethnography, including a long account of the enslaved people of Jamaica that sits in between a discussion of climate and another about horses, cattle, and dogs, perhaps indicating where in Sloane's mind enslaved people were ranked. He starts by describing enslaved Native Americans, and the neutral tone of the philosophical investigator almost immediately descends into value judgment. Indians, he says, "are very good Hunters, Fishers, or Fowlers, but are nought at working in the Fields or slavish Work, and if checkt or drub'd are good for nothing, therefore are very gently treated, and well fed." The final assertion seems unlikely but may have reflected his own narrow personal experience. The former part of the sentence advances a distinctly colonial ideology in that it both normalizes violence and assumes that the Native Americans on the island exist purely for the planters' benefit.[10]

Sloane's description of the island's enslaved Africans likewise begins with an assessment of the planters' conception of their value to the plantation economy rather than with impartial description. "The *Negros* are of several sorts," he begins, "from the several places of *Guinea,* which are reckoned the best Slaves. . . . Those who are *Creolians,* born in the Island, or taken from the *Spaniards,* are reckoned more worth than others in that they are season'd to the Island." From the outset, enslaved Africans are viewed as economic units rather than as individuals. Although there is a return to a more measured descriptive tone for a page or so, Sloane's utilitarian attitude has been clearly established. Unlike Ligon, his attitude is rarely balanced out by a sympathetic representation of enslaved people: Sloane chooses not to humanize enslaved people by telling their individual stories, nor to employ moralizing fables and parables to illustrate the human cost of slavery. Ligon had told a parable about a slave called Macaw learning to play the lute in which he had asserted the general point that some Africans are "capable of learning Arts." Sloane describes African music in more detail but, if he does draw any conclusion, it is merely that Africans are unable to control their passions. "The *Negros* are much given to Venery," he claims, "and although hard wrought, will at nights, or on Feast days Dance and Sing; their songs are all bawdy, and leading that way." The only vaguely sympathetic element of this is the recognition that enslaved people are "hard wrought," or overworked, but this is offset by the moralizing assertion that their songs are all bawdy. He does, however, take a technical interest in their music making. "They have several sorts of Instruments in imitation of Lutes," he notes, "made of small Gourds fitted with Necks, strung with Horse hairs, or the peeled stalks of climbing Plants or Withs. These Instruments are sometimes made of hollow'd Timber covered with Parchment or other Skin wetted, having a Bow for its Neck, the Strings ty'd longer or shorter as they would alter their sounds." This seems less judgmental than the description of the singing and dancing but, like Ligon, Sloane assumes that the instruments he observes are imitations of European originals. In fact, what he is describing could be any one of several traditional West African lute-like instruments, among which the best known is the kora. Kora players were often highly ranked members of African communities, and the instrument is associated with the Griot storytelling tradition. Sloane shares Ligon's ignorance of African musical culture and technology but fails to make up for it by seemingly lacking Ligon's ability to view the musicians as unique and creative individuals. Even when Sloane produces two pages of musical notation of African music—one of the earliest settings of African traditional music in Western notation—one gets little sense of the musicians who produced it.[11]

Sloane's description of the living conditions of enslaved Jamaicans ranges across a dozen pages and continues the pattern of presenting colonial ideology as neutral information, either by asserting as truths ideas about Africans that are clearly European opinions, or by presenting morally laudable behaviors neutrally while criticizing morally objectionable behaviors. For example, if cleanliness is next to godliness, we would expect the observation "The

Negros and *Indians* use to Bath themselves in fair water every day, as often as conveniently they can" to be presented positively. It is not, but neither is it presented negatively. It stands on its own as a paragraph of nineteen words without comment or judgment. By contrast, Sloane is more expansive in his censure of African irreligion. "The *Indians* and *Negros* have no manner of Religion," he believes. " 'Tis true they have several Ceremonies, as Dances, Playing, *&c.* but these for the most part are so far from being Acts of Adoration of a God, that they are for the most part mixt with a great deal of Bawdry and Lewdness." This cannot be regarded as simple description. Morally objectionable behavior is emphasized, wittingly or otherwise, because the judgment accords with colonial ideology and Sloane's religious sensibilities. By the end of these twelve pages, Sloane's alignment with the plantocracy is overt. On the face of it, it looks as if he has entirely abandoned any pretense of impartiality in favor of an apology for the tortures inflicted on enslaved Africans by European colonists. The passage concludes with an extended description of "The Punishments for Crimes of Slaves":

> For Negligence, they are usually whipt by the Overseers with Lance-Wood Switches, till they be bloody, and several of the Switches broken, being first tied up by their Hands in the Mill-Houses. Beating with *Manati* Straps is thought too cruel, and therefore prohibited by the Customs of the Country. The Cicatrices are visible on their Skins for ever after; and a Slave, the more he have of those, is the less valu'd.
>
> After they are whip'd till they are Raw, some put on their Skins Pepper and Salt to make them smart; at other times their Masters will drop melted Wax on their Skins, and use several very exquisite Torments. These Punishments are sometimes merited by the Blacks, who are a very perverse Generation of People, and though they appear harsh, yet are scarce equal to some of their Crimes, and inferior to what Punishments other European Nations inflict on their Slaves in the East-Indies.

This passage has been widely discussed. Late eighteenth-century abolitionists such as Anthony Benezet and Thomas Clarkson saw in it evidence of the enormous cruelties inflicted by slaveholders while modern critics have considered it an early statement of conscious antislavery, revealing the true nature of slavery to an unwitting metropolitan audience. James Delbourgo has, however, pointed out that in later discussion the passage "became what it had not been before: part of a moral and political attack on the unenlightened 'cruelties' of slavery, and an argument for abolition." In fact, argues Delbourgo, the passage is "neither anti-slavery nor pro-slavery, because the moral economy of curiosity demanded no such stance." As Delbourgo points out, strictly speaking Sloane was neither for nor against slavery because those firm positions were rarely taken at the point at which this was written. Nevertheless, Delbourgo argues that critics have ignored some important aspects of the passage, such as Sloane's contention that "these Punishments are sometimes merited by the Blacks" (which is hardly an overt statement of antislavery) while overlooking the ambiguity inherent in the word "sometimes." In his later biography of Sloane, Delbourgo argues more firmly that Sloane was a defender of slavery in this passage which, he argues, "is not surprising. By the time he published the *Natural History* in 1707, he was not a mere observer of the institution, but an interested participant in its workings with a significant financial stake in the plantation system through his marriage." Either way, Sloane's writing tends to normalize slavery even when it takes no clear stand. His seemingly impartial descriptions could be, and were, used as ammunition by later abolitionists. On closer examination, however, his language is often less than neutral and more closely allied with the interests of Jamaica's planters and he represents even the most appalling tortures as an integral and accepted part of the plantation system.[12]

The passage presents supposed evidence of the proportionate, reasonable nature of the colonists. Sloane assures the reader that "Beating with *Manati* Straps is thought too cruel, and therefore prohibited by the Customs of the Country." Delbourgo notes that Sloane had a manatee-hide whip in his cabinet of curiosities, given to him by "John Covel, master of Christ's College, Cambridge." He concludes that the whip may have reminded Sloane of the extravagance of Caribbean punishment: "a literal braiding together of the extremity of the social relationships being forged in the West Indies" (although in his later biography Delbourgo concludes that Sloane "defended the tortures he described"). If such punishments were prohibited, however, one wonders how Covel obtained the illicit weapon, or why anyone would have made one in the first place. On closer reading, we see that it is merely the custom of the country, not the law of the land, that prohibits use of the manatee-hide whip. Indeed, Sloane even presents evidence of its use, and reports on the devastating impact it has both on the body and on the perceived economic value of the victim. "The Cicatrices are visible on their Skins for ever after," he tells us, "and a Slave, the more he have of those, is the less valu'd." This discussion of the manatee-hide whip confirms not that it was prohibited but, conversely, that it was actually used. Sloane's assessment of its effect on the value of slaves reinforces only that enslaved people are treated as chattel rather than as human beings. The effect, once more, is to normalize the brutal treatment of enslaved people, to suggest that their punishment is subject to reasonable checks and balances and palliated by the customs of the country.[13]

The theme is extended and concluded in the final paragraph. Here, the tortures that Sloane describes are not only truly horrendous but also wantonly sadistic: salt and pepper rubbed into lacerations and melted wax poured onto the skin. Sloane calls these "very exquisite Torments," which might suggest that he thinks them excessive, but what follows is ambiguous. He argues that "These Punishments are sometimes merited by the Blacks, who are a very perverse Generation of People." The "sometimes" can go both ways, suggests Delbourgo, meaning that Sloane believes some of these punishments are merited while some are not. Sloane calls Africans "a very perverse Generation of People," implying that they are perfectly capable, in his view, of behaviors that merit exquisite torments. The "sometimes" refers to the behavior, which is occasional, rather than to the punishment, which is standard. Indeed, Sloane goes on to claim that "though they appear harsh," these punishments are if anything lenient, being "scarce equal to some of their Crimes." The "some" here mirrors the "sometimes" in the previous sentence, reinforcing that when the crime is severe enough, Sloane believes the tortures to be proportionate. Yet again, Sloane normalizes torture, arguing that the punishment is appropriate when it fits the crime. The passage concludes with another claim about the moderate and relatively lenient nature of these torments. They are, he says, "inferior to what Punishments other European Nations inflict on their Slaves in the East-Indies." This is a classic red herring, drawing attention away from one abuse by pointing out another elsewhere. Taken individually, Sloane's descriptions of slave punishments in this passage could be used as evidence for the cruelty of slavery. Taken as a whole, the passage reveals that Sloane was well able to justify slavery and the severe treatment of enslaved people.

Sloane's detailed description of enslaved people in Jamaica ends here, although there are occasional references to "Negros" throughout the rest of the introduction. Sometimes they are represented as enslaved, sometimes as maroons, but most often as Sloane's patients. He describes yaws, worms, one case of apparent depression (which he calls "madness"), and a couple of cases of what he believes to be Africans feigning illness. In general, these mentions remain descriptive and avoid moral judgment, but this is again a normalizing approach. Unusually, Sloane gives the name of the enslaved woman who was suffering from depression—she was called Rose—and he speculates on the causes, noting that her episodes coincided with full moons, but either it does not occur to him that her enslavement may have been a contributory factor or he suppresses that insight. He condemns traditional African medicine as ineffective. "There are many such *Indian* and Black Doctors, who pretend, and are supposed to understand, and cure several Distempers." On examination, however, he decided that "they do not perform what they pretend, unless in the vertues of some few Simples. Their ignorance of Anatomy, Diseases, Method, *&c.* renders even that knowledge of the vertues of Herbs, not only useless, but even sometimes hurtful to those who imploy them." By contrast, he insists on the good medical care afforded to enslaved people by European colonists, since "Planters give a great deal of Money for good Servants, both black and white, and take great care of them for that Reason, when they come to be in danger of being disabled or of Death." Once again, this paints a rosy picture of plantation life, but not one that is outside the bounds of normal experience, and it tends to reassure the reading public in England that all is well on the nation's carefully managed plantations.[14]

Sloane's medical accounts provide a fascinating insight into the daily minutiae of life in a tropical colony, but they are methodologically somewhat removed from natural history, the book's main subject area. After fifty pages describing the voyage, the remaining six hundred pages of the book catalogue and describe Jamaica's wildlife, cultivated plants, and domesticated animals. This catalogue is peppered with frequent, but passing, references to enslaved people that confirm they are primarily associated with economically important species. Despite Sloane's intention to seek "the best Informations I could get from Books, and the Inhabitants, either *Europeans, Indians,* or *Blacks,*" the chapters describing such economically marginal organisms as seashore species, fungi, and ferns rely on books and Europeans alone. By contrast, Sloane provides a long history of maize (*Zea mays*) and how it is grown by Indians, telling us that a species of rice (*Oryza sp.*) "is sowed by some of the *Negro*'s in their Gardens," and that castor oil (from the castor oil plant, *Ricinus communis*) is "good for clearing *Negroes* Skins, and for Lice on the Head." Curiously, he does not mention the involvement of enslaved people in growing sugarcane. As the castor oil example suggests, many of the mentions of enslaved people are medical in nature. Sloane's *Voyage* was not merely an abstract scientific treatise; it was also intended as an essential tool for physicians and apothecaries, both those tending to European colonists and those responsible for enslaved Africans. Both European and African remedies are discussed, although the former predominate. Among those involving Africans, Sloane recalls how he "saw once a Black stop a bleeding artery" with Dutch grass (probably *Eleusine indica* stems used as a tourniquet), that "however rude the Labour or Travel in Childbed of the Savages is," the powdered root of nutgrass (*Cyperus rotundus*) "taken in White-Wine, makes them be speedily delivered," and that long pepper (*Piper longum*) is "thought to ease pain in every affected part, and therefore this is esteemed as a very rare Remedy, by all Indians and Negroes, and most part of Planters, but I could not find that this Leaf could do any more than Coleworts" (probably wood avens, *Geum urbanum*.) These are just a few examples. What they tell us is that, for Sloane, enslaved people were an essential, but not exceptional, feature of the Jamaican landscape. Slavery was simply a routine part of the business of the island, fully naturalized to the Caribbean environment.[15]

In 1725, Sloane published the long-awaited second volume of the *Voyage*. Most of this continued and completed his catalogue of Jamaican species. As before, enslaved people are occasionally mentioned as figures in a colonial landscape but, although the pattern is essentially the same, one passage is worth reproducing in full. Anxious, or proud, to prove both the utility and the accuracy of his botanical descriptions, Sloane includes a short anecdote, related to him by the naturalist Henry Barham (c. 1670–1726) "in a Manuscript, call'd Hortus Americanus," which illustrates how useful his work had become in the field. Barham

> took notice to me of an accident, whereby several Negros had been poyson'd in the year 1711. The Account he gave me was this, that a Negro Servant carrying some Rum in a Vessel upon his Head, as their Way is, found, that upon motion, it run over, to stop which he pluck'd the Leaves of a Plant he found growing in the Savanna or Meadow, over which he was passing. Upon drinking this Rum they found the Negros poyson'd, some whereof I think died, and thereupon the Negro was try'd for his Life, the rest recover'd by the Use of the Juice of the Indian Arrow Root, or Canna Indica radice alba Alexipharmaca, of my Catalogue of *Jamaica* Plants. p. 122. Hist. Vol. I. p. 253. Mr. Barham observing these Leaves, and comparing them with my Description and Figure, found them presently to be of the Apocynum erectum fruticosum flore luteo maximo & speciosissimo, Cat. p. 89. Hist. Vol. I. p. 206.

The poisonous plant was probably dogbane (*Apocynum cannabinum*), a common invasive and poisonous weed, while the cure was almost certainly West Indian arrowroot (*Maranta arundinacea*). What strikes the modern reader most forcibly, however, is not the botanical details but the unresolved human story. We do not learn if the "Negro Servant" who was put on trial was acquitted or convicted. Sloane himself appears to lose interest in the human being behind the story once the utility of his natural history is proved. Again, an enslaved African is represented as a passive figure in the landscape. His actions are ostensibly described faithfully, without either praise, criticism, or sympathy. We note, however, that his appearance as a mere narrative device in a bigger story mirrors the planters' exploitation of individual Africans as part of a wider imperial and economic project. This "Negro Servant's" story involves life and death decisions and invites a complex response from readers learning about a capital trial for what must surely have been an honest mistake. Astonishingly, however, Sloane does not assess this individual's misfortune by any moral, legal, or political criteria, nor does he see him as anything else than a naturalized element of an ordinary plantation.[16]

Sloane's *Voyage* marks an important turning point. Emerging from tentative and sometimes less than coherent seventeenth-century accounts of slavery in the Caribbean environment, it set the bar for naturalists for the following century. Its publication was the moment at which ethnographic description of enslaved people in Caribbean natural histories accepted slavery as naturalized to the colonies, thereby sweeping aside any residual anxiety or ambiguity that older natural historians like Ligon might have expressed about the morality of slavery. Sloane's certainty was accepted and adopted by those who wrote in the service of the plantocracy throughout the first half of the eighteenth century such that, by 1760, few if any naturalists or plantation manualists presented slavery as anything other than a natural phenomenon. The monolithic certainty that Sloane engendered was, however, far from stable. The strength of the assimilation of slavery to nature in early eighteenth-century scientific writing gave abolitionists of the late century a firm base from which to attack the slave trade and even slavery itself as unnatural. Thus Sloane's *Voyage,* which was for more than fifty years the planters' bulwark, ultimately became a powerful weapon in the abolitionists' arsenal.

GRIFFITH HUGHES AND THE NATURAL HISTORY OF BARBADOS

Griffith Hughes's 1750 *Natural History of Barbados* presents itself as a landmark of formal natural history, self-consciously in the tradition of Pliny, Ray, and Sloane. A later generation saw it as an early antislavery text. Hughes is the earliest author discussed in this book to be included in Thomas Clarkson's famous map of "Abolitionist Forerunners" (see figure 1), published in 1808, because of the way Hughes "took an opportunity . . . of laying open to the world the miserable situation of poor Africans, and the waste of them by hard labour and other cruel means." Hughes himself remains something of an enigma, appearing in contrasting roles at different times and places. He became for a while a minor scientific celebrity, and then disappeared without trace. He was born in Tywyn on the Welsh coast in 1707 and educated at Oxford in the late 1720s before being sent to Radnor, Pennsylvania, by the Society for the Propagation of the Gospel where he became the rector of St. David's Church. He preached in both English and Welsh, adopting the character of a rugged backwoods preacher, sometimes walking many miles to neighboring parishes and preaching under the shade of large trees. Unexpectedly, in 1736, he abandoned his parish, complaining of a bad knee, and accepted the living of St. Lucy in the scenic but relatively remote northern tip of Barbados. Here he reinvented himself as a parson-naturalist, paying so much attention to his scientific research that his new parishioners complained that he neglected them. He visited London at least twice over the next fifteen years. These visits must have taken him away from the island for several months, at the very least, and the legend persists in Barbados that he took the parish records with him on one such trip and forgot to return them (they are still missing). In London, he established himself first as a scientist, gaining election to the Royal Society in 1748, and next as a scientific writer. His *Natural History* was published in 1750 with an extraordinarily extensive list of almost five hundred subscribers, which included several of the crowned heads of Europe, dukes, earls, barons, and bishops, as well as most of the scientific establishment of the day, including the elderly Hans Sloane. Thereafter, Hughes vanishes from the record, although there are contradictory reports of him dying in London in 1758 or returning to Barbados and surviving into the 1760s.[17]

Hughes's *Natural History* is a large and handsome folio volume, lavishly illustrated, and somewhat resembling Sloane's *Voyage to Jamaica* in its appearance and its use of John Ray's method. The book was soon attacked for its scientific weakness and inaccuracy. John Hill (c. 1714–75) subjects it to a ten-page savaging in the *Monthly Review* that veers between scientific impartiality and open satire—a reflection, in fact, of the reviewer's own career. Hill was an actor and satirical writer as well as a competent botanist. Spurned in his attempts to join the Royal Society, he was clearly galled that Hughes had been admitted on such easy terms. His chief objections, among many in his compendious drubbing, are that Hughes has not followed an adequate scientific

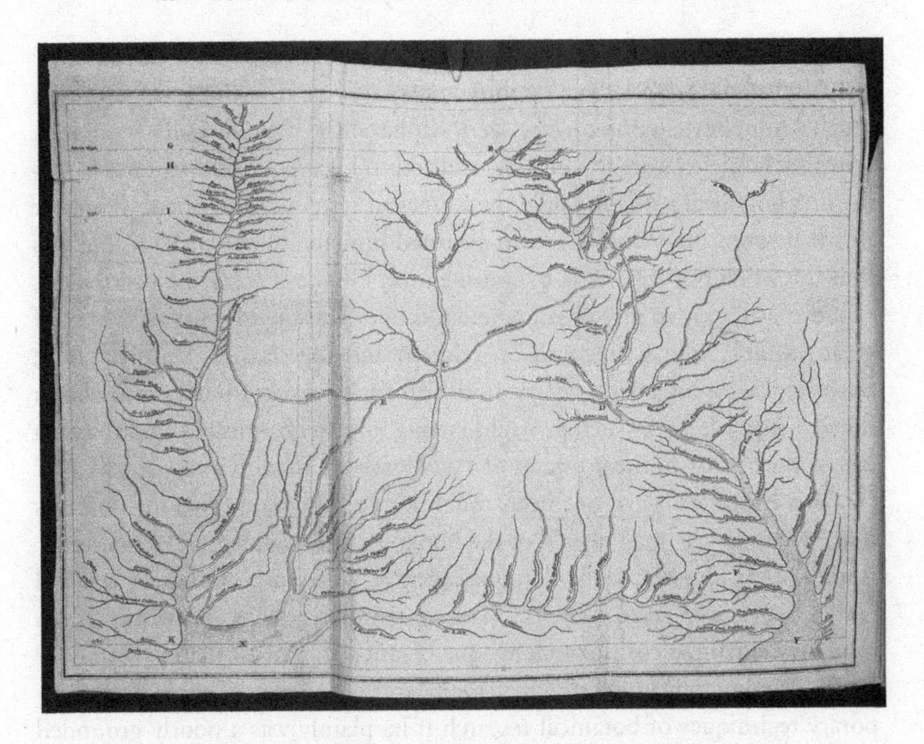

Figure 1. Foldout map depicting abolitionist forerunners as a system of rivers and tributaries. From Thomas Clarkson, *The History of the Rise, Progress, and Accomplishment of the Abolition of the African Slave Trade,* 1808. (© The British Library Board.)

method and that he has relied too much on local Barbadian names without giving precise scientific names:

> What idea can any one form of a tree from its being called *The Anchovy-Apple, The Bread and Cheese,* or *Sucking-bottle? The Maidenhead, The Belly-ach, The Spirit-weed, The Cat's-blood, The Bumbo-bush, The Hogslip, The Fat-pork Tree, The Christmas-bush, The Gully-root, The Sweet-heart, The Wild Dolly, The Cuckold's Encrease, The Pudding-wyth, The Dumb-cane,* and a multitude of others that might be cited, are of the same turn. They may serve, indeed, in the language of a planter, to distinguish objects daily before them from one another, and to put them in mind of the more essential blessings of fat pork and pudding; but what information can be convey'd to people here under such names (and none of these have any other,) we are at a loss to guess.

One detects a certain metropolitan snobbishness about both the planters' botanical abilities and their famous appetites, but it is certainly true that Hughes supplies scientific names very haphazardly, if at all. Hill's interest in scientific nomenclature was current and active; his *Flora Britanica* of 1759 was the first botanical study to apply the Linnaean system to the British flora, for which in 1774 the Swedish crown awarded him the Order of Vasa. He was thus the worst possible critic to comment on Hughes's *History.* Nevertheless, Hill tries valiantly to give credit where it is due, praising Hughes as "a man of great probity" who writes well on religion and the classics. Unfortunately, however, "as to his talents for natural history, it was an unlucky mistake in him, to suppose them such as could enable him to go through so arduous a task as the history of the products of a whole island, though a very small one." Modern historians have not been much kinder. Raymond Stearns observes that "whatever else may be said of the book, it was a masterpiece of salesmanship. . . . Mr Hughes, while demonstrating a wide grasp of ancient classical works, appears to have had only passing familiarity with the works of his more immediate scientific predecessors." Francis. J. Dallett damns with faint praise, arguing that Hughes "was actually entirely abreast of general contemporary techniques of botanical research if he plainly was a poorly grounded scientist and not prepared to do a thorough job."[18]

It is easy to find examples of the book's shortcomings. Serious scientists looking in this expensive publication for detailed morphological description and a clear identification, with precise measurements and a Linnaean Latin name, might have felt short-changed when they read the following three entries, here given in their entirety:

> *The* SMALL YELLOW-BIRD
> THIS is a very small Bird, whose Plumage hath a beautiful Mixture of Yellow and Red, especially about the Head.
>
> *The* BLACK MAIDEN-HAIR
> I Found this beautiful Plant growing upon the Wall of *St. Lucy*'s Church.
>
> *The* DOMINICO-LOBSTER
> THIS Lobster seldom weighs more than about two Pounds.[19]

The book's deficiencies as natural history might be all too apparent, but the work was nonetheless successful in perpetuating the notion of slavery as an integral and unshakable part of the Caribbean social and environmental landscape, and it would be frequently quoted by both defenders and opponents of slavery. As befits a work by a clerical naturalist, its preface offers a justification for its own existence expressed as detailed statement of natural theology. It is through contemplating the "inexhaustible Variety" of the world, argues Hughes, that "we are gradually led from Things visible, to the Knowledge of him who is invisible. . . . By these, and such-like Inquiries, we find in every thing a wise, good, and useful Design." Like Sloane before him, Hughes is in almost all cases a reductionist thinker, rarely discussing the interactions between species other than to consider for what purpose God had created them. One of the few places where he begins to think about interactions is in his discussion of flying fish, the animal that had prompted Richard Ligon to conceive of the ocean habitat in a way that we might today think of as proto-ecological. Like Ligon and Sloane, Hughes also thinks about the fish's ability to escape from dolphinfish by flying into the air and he also grasps the basics of population dynamics by arguing that the fishes' "increase is prodigiously great; otherwise the whole Species must have long ago been destroyed; for they are Prey to Men, Fish, and Birds." He goes on to think in more detail about predator-prey interaction, noting that if, to avoid the dolphinfish the flying fish "seek Refuge in the Air, a Bird called the *Cobler*, among a great many others, darts with the Swiftness of an Eagle to destroy them." "Cobbler" is a Barbados dialect word for the magnificent frigatebird, the main predator of flying fish, although metropolitan readers would be forgiven for not realizing that since, in another example of his haphazard method, the species is not mentioned in his own chapter on birds. Nevertheless, this entry comparing a prey species and two of its hunters is very much the exception. Almost every other species is described in isolation, found in the position where Hughes believes that God had placed it.[20]

Like Ligon and Sloane, and in the tradition of Pliny, Hughes offers an ethnographic account of the enslaved population of Barbados early in his book as part of his description of the island's physical and social geography. Unlike the other naturalists, he does not supply even one enslaved person with a name, and he uses the words "slave" and "Negro" interchangeably, suggesting that, in his mind, the two go naturally together. Most of his mentions of African people are brief, simply mentioning their use of various plants or animals. Some are purely descriptive, such as that Africans eat various vegetables or adorn themselves with various decorative natural objects. Not all mentions are neutral, however. Of those that make or imply a judgment, none are positive and about a dozen are derogatory, showing absurd, contemptible, or supposedly amusing behavior.

Thus he blames a natural phenomenon, a wildfire in a bituminous outcrop, on "a Slave roasting Potatoes upon the side of an Hill." A few pages on, in his discussion of the Hag's Horse (a stick insect, the godhorse *Paraphanocles keratosqueleton*), he is amused that "a great many Negroes have a Notion, that, if they kill one of these, they will be very unlucky in breaking all Earthen Wares they handle." He is later equally amused by "a remarkable Instance of Superstition of a Negro" who believed that his rheumatism was caused by him killing a snake, which took supernatural revenge. These examples are patronizing, portraying Africans as innocent and primitive, but harmless. His account of a "certain Slave" who attempts to poison her master with "manchaneel" (the notoriously poisonous manchineel tree, *Hippomane mancinella*) is more judgmental. "Conceiving herself injuriously treated, [she] poured into her Master's Chocolate about a Spoonful of this juice." Although seriously ill, he survived, which Hughes speculates might have been due to the chocolate working as an antidote.[21]

These anecdotes illustrate the nature of the plant or animal under discussion while revealing Hughes's conception of African nature. They serve Hughes's immediate purposes, just as enslaved people served the purposes of the planters, but they perpetuate colonial attitudes and represent enslaved people as integral to the colonial project. Hughes is also inconsistent. The single longest botanical entry in the book is that for sugarcane, which takes up nine pages, but Hughes almost completely elides the existence of the enslaved workers who grew and refined the cane. The only mentions are a brief footnote and an account of an incident in which three enslaved people died after being overcome with fumes when trying to clean out a rum storage cistern. A fourth worker, "a white Person, who was a Workman on the Estate," survived only because he took a rope with him into the cistern as a lifeline, demonstrating that health and safety concerns were taken more seriously when white workers were at risk. Otherwise, one would get the impression from Hughes's account of the culture and refining of sugarcane that African labor was not involved.[22]

The main discussion of the island's enslaved population takes place early in the first section of the book. For a moment, Hughes looks as though he is going to bemoan the worst excesses of plantation slavery, complaining that "Children, in these *West-India* Islands, are, from their Infancy, waited upon by Numbers of Slaves." This tends to give the children "an overfond and self-sufficient Opinion of their own Abilities, and so become impatient, as well as regardless, of the Advice of others." This was a frequently made observation at the time, but Hughes's intervention is less a critique of slavery than a remark on parenting among the settlers. It is followed up by five ethnographic pages that include description of the physical appearance and cultural practices of Africans in Barbados, before launching into a conventional defense of the slave trade. Like Sloane, Hughes strives to maintain a balanced tone, but his sympathies are clearly with the plantocracy. For example, his ethnography contains a statement that is deeply, if unintentionally, ironic. "The Negroes," he observes, "are very tenaciously addicted to the Rites, Ceremonies, and Superstitions of their own Countries, particularly in their Plays, Dances, Music, Marriages, and Burials. And even such as are born and bred up here, cannot be intirely weaned from these Customs." As an Anglican priest, Hughes's main duties were to administer the rites and ceremonies of a church from a country 6,500 kilometers (4,000 miles) away. The passage reminds us how deeply internalized colonial attitudes could be, but also that for Hughes both Anglican religious culture and the plantation system to which it ministered were naturalized to Barbados. Hughes follows this passage with a description of African customs and beliefs, including a long footnote on Obeah in which he maintains that "Obeah Negroes," whom he dismisses as "conjurers," get "a good Livelihood by the Folly and Ignorance of the rest of the Negroes." Such descriptions reinforce the attitude present throughout the book that African culture is primitive and Africans themselves gullible and impressionable.[23]

We should note, however, that Hughes does not use either the culture of Africans or their physical attributes as a direct justification for their enslavement. Drawing on other anatomists of the period, he dismisses climatic explanations for racial difference, arguing that "as to the Blackness of the Negroes Skin, this reaches no deeper than the outward *Cutis,*" and that "as to the Stature and Make of Negroes, excepting that a greater Number of them have their Noses shorter, and Lips thinker, than the *Whites,* I never could find out any extraordinary Difference." Moreover, "the Capacities of their Minds in the common Affairs of Life are but little inferior, if at all, to those of the *Europeans.* If they fail in some Arts, it may be owing more to their Want of Education, and the Depression of their Spirits by Slavery, than to any Want of natural Abilities." Such arguments would become a staple of abolitionist literature just a few decades later. Here, they are presented as descriptive natural history with no direct bearing on Hughes's views of the slave trade. His frequent follow-ups with comparisons between modern Caribbean slavery and classical and biblical slavery, however, remind us that he not only saw slavery as natural but also as ordained. At any given moment in history, some group or nation of people were enslaved by others. Currently, it was the turn of Europeans to enslave Africans.[24]

In the second half of his discussion of Africans in Barbados, Hughes announces, "Since I have made this Digression to treat of the Manners and Customs of these Negroes, it may perhaps be expected, that I should consider the several Arguments for and against making our Fellow-creatures Slaves." It is interesting that he uses the word "expected," since the debate over the legality of slavery did not take off in any sustained way until later in the eighteenth century. As a clergyman, Hughes might have considered it important to take a position in the moral or spiritual spheres, although he does explicitly shy away from "engaging in a Controversy foreign to my Design." It is possible that during his time in Pennsylvania in the 1730s, Hughes had come into contact with Quakers such as Benjamin Lay (1682–1759) who were at that time agitating to forbid Friends from keeping slaves. More likely, Hughes is responding to a generally submerged but increasingly insistent sense across the English-speaking world that slavery needed to be justified, at least, and may not ultimately be justifiable. Whatever brought Hughes to this, his position rapidly becomes clear: "The Power of making Slaves is, and hath been, a natural Consequence of Captivity in War." Again, he sees this as "natural." The idea that war might be fought differently, or even not fought at all, does not occur to him. He argues that, well before Europeans arrived in Africa, "the several different Nations were so very savage and barbarous, that they were at continual Wars with one another." Prisoners were put to death and their dismembered limbs hung up in the trees. How much better that they are brought to Barbados, because of "the Benefit that arises to Mankind from their Labour" and because "at least a few, among many Thousands imported, may probably come to a better Knowledge of their Duty to God and Man." These are standard arguments in the planters' repertoire—Hughes is saying nothing original. His conclusion is simply an appeal to his readers "gratefully to acknowledge the Happiness of living where neither our Lives nor Fortunes are at the Mercy of any tyrannical Oppressor." Once again, this reinforces slavery as part of God's natural order, and Hughes asks his readers to question neither its utility nor its legality, but instead only to reflect on their own good fortune and the happiness of their nation.[25]

"From the Disposition and Manners of the Inhabitants," Hughes tells us, "the next Thing that will naturally fall under our Consideration, will be the Nature of the Soil." Why Hughes thought this would "naturally" follow is not clear. Although Pliny's ethnography appears in book 7, long before his discussion of soil in book 18, on agriculture, by the early modern period most naturalists began with soil, which was considered the lowest rung on the great chain of being—Ligon and Sloane both considered soil before they described people. Hughes's "naturally" offers a clue to his underlying thinking. He describes the soil in low places as black, in "shallower parts" as reddish, and on top of the hills as white. The comparison with the racialized social structure that he had described immediately beforehand is obvious, although one assumes that this was an unintentional rather than a deliberate rhetorical strategy. When he goes on to proclaim that "by such variety, Providence hath wisely adapted different Soils to the different Nature of the several Kinds of Trees, Shrubs, and Plants," one cannot avoid concluding that, subconsciously at least, Hughes believes that just as God adapted soils to fit the trees necessary to the colonists, so he adapted enslaved people to the tasks necessary to colonists, especially when Hughes starts to discuss the black soil as being most suitable to support the sugarcane. Of course, sugarcane does indeed require a dark soil high in organic content, but Hughes's organization of the information is revealing. The people of the island, free or enslaved, are there as part of a natural hierarchical system ordered by God for the benefit of the colonists. Although contemporary readers derided *The Natural History of Barbados* for its defective science, in this opinion Hughes was very much in the mainstream of both colonial and scientific thought. Emulating the formal set-piece natural history of Sloane, Hughes added yet more weight to the by now conventional view that slavery was fully naturalized to both the environment and the society of the British colonies in the Caribbean. Clarkson's inclusion of Hughes among the abolitionist "forerunners" appears, therefore, to have been generous at best.[26]

Whether deliberately or inadvertently, and despite their somewhat variable attempts at scientific impartiality, Sloane and Hughes produced natural histories that both represented and perpetuated plantation attitudes. These works were, however, among the few readily accessible and intellectually substantial sources available to readers interested in finding out more about plantation slavery in the decades before abolitionists flooded the literary marketplace with damning representations of colonial enslavement and the slave trade. Ironically, texts written in the service of the plantation would become central documents of the archive drawn upon by abolitionists and, in their turn, important pieces of evidence in the case against the slave trade and plantation slavery.

CHAPTER 3

"Negroes, cattle, mules, and horses"

THE PLANTATION IN THEORY AND IN PRACTICE

Mid-eighteenth-century readers interested in plantation slavery could find information scattered across a variety of different types of writing: from voyage narratives to religious, economic, and legal treatises. They were most likely to find what they were looking for, however, in the many texts that were principally concerned with Caribbean nature and landforms which, as well as more conventionally scientific accounts of flora and fauna, included agricultural treatises, locodescriptive and georgic poetry, and island histories. Together, these texts became an important addition to the archive of colonial environmental writing. Most were, however, explicitly or implicitly written in the service of the plantocracy, and they tended to represent enslaved people as little more than economic units laboring for a plantation system that was fully naturalized to the region. Despite this bias, and in the absence of a substantial alternative literature, they were unavoidably a major source for later authors, including both abolitionists and proslavery apologists, who wanted to bolster their arguments with firsthand accounts and the testimony of experts. Caribbean environmental writing of the mid-eighteenth century thus had a significant impact on the tone and content of the late eighteenth-century abolition debate.

This chapter examines three environmental texts that, while not primarily natural histories, attempted to theorize mid-eighteenth-century plantations both as places where natural resources were accessed and as sites of agricultural labor. Between them, they added substantially to public discourses concerning the place of the plantation and its inhabitants in the natural environment. Samuel Martin's *Essay upon Plantership* (1750) was the century's most substantial guide to the theory and practice of plantation management, James Grainger's *The Sugar-Cane* (1764) poeticized the plantation in an attempt to situate it in an agrarian cultural tradition that dated back to the classical period, and Edward Long's *History of Jamaica* (1774) was a popular social, political, and natural history of Jamaica that became notorious for its contribution to the development of racist ideologies. While Martin's *Essay* is rarely considered today, Grainger's *Sugar-Cane* and Long's *History* have each received substantial scholarly attention in recent years, albeit if only with the aim of exposing their colonial attitudes. Cristobal Silva's observation that "there has been a robust and growing body of scholarship about Grainger and his work, though this revival has admittedly done little to burnish his reputation as a poet," applies equally to Long's reputation as a historian.[1]

Martin, Grainger, and Long shared a broadly ameliorationist outlook in that they accepted the Caribbean slavery system as a natural, or at least as a naturalized, phenomenon while appearing to critique its worst excesses. Although such attitudes were not uncommon in print in this period, and in some cases may even have reflected genuine attempts to change plantation practices, they were often merely decorous, were usually deliberately misleading, and even when sincere were always underpinned by the economic argument that a supposedly "reformed" plantation was more productive and thus more profitable. Thus, ameliorationist texts, although sometimes understood as representing a "third position," were in fact contributions to a burgeoning proslavery literature that sought to undermine the increasingly audible arguments of those who claimed that slavery was indefensible. Nevertheless, authors of ameliorationist literature had no choice but to allude to or even describe the plantation brutality that they aimed to reform and, for that reason, their writing could be as valuable to abolitionists seeking to expose cruelty as it was to apologists seeking to show how planters supposedly ameliorated brutality. Martin, Grainger, and Long were by no means the only ameliorative authors in this period, but they were among the most influential. Examining their work reveals the extent to which plantation slavery was by this period understood to be natural and inevitable by its defenders while also allowing us to understand the raw materials with which abolitionist authors were obliged to work.

A RIGHT OR RATIONAL METHOD? MARTIN'S ESSAY UPON PLANTERSHIP

Pirates and privateers may have gone to the Caribbean in search of booty, and natural historians in search of knowledge, but from the outset the majority of British settlers went as planters, a word that from the thirteenth century had been synonymous with "farmer" but that from the sixteenth century took on the new meaning of "colonist." The climate and the crops were new, but British planters were nonetheless farmers and businessmen, primarily concerned with making money from the land. The new environment required, however, that they discover novel farming methods and quickly pass on their successes and failures to others. Most agricultural knowledge in the early modern period was transmitted orally, with both landowners and tenants learning from previous generations. British farmers nevertheless had a long and expanding tradition of agricultural manuals to which they could turn, and these constituted a sort of parallel discourse to natural history, offering a more practical and, some would say, a more instrumental approach to understanding the environment. Agricultural manuals dated back to the classical period, with Cato the Elder (234–149 BCE), Varro (116–27 BCE), and Columella (4—c. 70 CE) offering farming advice alongside advice on managing enslaved laborers, with a similar mix appearing in the more aestheticized approach to agriculture offered in poetry, notably Virgil's (70–19 BCE) *Georgics.* In the Middle Ages, sumptuously illuminated manuscript books offered advice on the farming year to the minority able to consult them, but with the advent of printing, agricultural manuals, almanacs, and garden calendars began to proliferate. The trend was set by Thomas Hill (c. 1528—c. 1574), whose gardening manuals became popular in Tudor and Stuart Britain, but other countries developed their own traditions suited to their particular climates. In France, for example, Olivier de Serres (1539–1619) offered advice on viticulture as well as agriculture in his *Théâtre d'agriculture* (1600). By the late eighteenth century, as James Fisher has noted, agricultural manuals had become so widely available that those who set out to farm their land based on reading rather than hands-on experience became known contemptuously as "book-farmers," mocked "for reading too many 'farming romances' and living out mad fantasies as 'knights-errant in farming.' "[2]

The first British settlers in the Caribbean had no such tradition to guide their clearing and planting. To start with, they proceeded by trial and error and by gleaning what they could from manuals rooted in rural England. As Jennifer Mylander has pointed out in a study of colonial manual ownership in Virginia, these manuals were often highly impractical in utilitarian terms but served an ideological function "for colonists committed to identifying themselves not as Americans, but as the *English* in America." This may also have been true in the Caribbean, but English manuals would have been even more impractical in tropical Barbados than in temperate Virginia. Accordingly, some planters began to write their own manuals. These authors took a pragmatic view of both agriculture and the management of enslaved people and passed on their knowledge of how to run a plantation, usually based on personal experience. Sometimes that personal experience impelled them to recommend humane treatment of slaves, while at other times they justified extreme punishments. In either case, their pamphlets normalized an unequal relationship in which advice about management of enslaved people more closely resembled guidance on animal husbandry than on managing human workers.[3]

Richard Ligon had offered advice to planters, but his suggestions and techniques belonged to the early phase of British colonization in the Caribbean, as did his disquiet about slavery. As the colonies became better established, the planters became more experienced at their craft and more entrenched in their views. Just as natural historians of the mid-eighteenth century had naturalized slavery to the Caribbean environment, so the authors of plantation management manuals normalized the management of enslaved people. Planters such as Samuel Martin offered practical advice to planters but less often debated the rights or wrongs of slavery, even though they sometimes presented themselves as benevolent and caring slaveholders. Where natural historians such as Griffith Hughes felt it necessary to outline the reasons why slavery could be defended, plantation managers usually took it for granted that slavery was simply the normal and natural system of the colonies. They often advocated what they considered to be reasonable and humane management of enslaved people, but their justifications were invariably more concerned with profit than the well-being of the people whose labor they commanded.

Samuel Martin (c. 1694–1776) was at the center of Caribbean society and politics for much of the eighteenth century. His *Essay upon Plantership,* published in Antigua in 1750, was a popular and important introduction to the topic; indeed, Nathalie Zacek has claimed that it was "the most widely read tract among West Indian colonists on the subject of plantation agriculture." Martin was the grandson of George Martin (fl. 1650–70), a Royalist exiled to Surinam by Cromwell who appears as a minor character in Aphra Behn's fictional account of Oroonoko's uprising in Surinam, and the son of a planter, also named Samuel (d. 1701), who had been killed by enslaved people in an uprising in Antigua at the start of the century. Educated at Cambridge, the younger Martin was elected to the Antigua Assembly in 1716, becoming its speaker after 1750, and colonel of the island's militia. In the final three decades of his life, he dedicated himself to reconstructing his plantation, Green Castle, on lines that some historians have called "reformed," "innovative," or even "enlightened." This was certainly the impression he sought to foster, as the Scottish traveler Janet Schaw, who thought Martin "the most delightful character I have ever yet met with," amply attests. Schaw, who visited the elderly Martin late in 1774, noted in her journal that his estates "are cultivated to the height by a large troop of healthy Negroes, who cheerfully perform the labour imposed on them by a kind and beneficent Master, not a harsh and unreasonable Tyrant. Well fed, well supported, they appear the subjects of a good prince, not the slaves of a planter." Martin may well not have been the worst planter on the island, but as Schaw observes, this was primarily a business rather than a humanitarian policy: "The effect of this kindness," she remarks, "is a daily increase of riches by the slaves born to him on his own plantation." Whether "reformed" or not, at no point did Martin advocate an end to slavery. He also become a prolific writer, leaving his influential *Essay upon Plantership* as well as more than twelve hundred letters, not yet adequately studied, that reveal a man with numerous global connections. The *Essay* went through seven editions in his lifetime and several reprints thereafter, with editions published after 1773 carrying an additional chapter that explicitly challenged the "zealous enthusiasts for the freedom of negroes" who had started to make a mark on public discourse in the wake of the Somerset case of 1771–72, which effectively outlawed slavery in England. The book was influential, probably because many erroneously believed, as did the apologist for slavery Gordon Turnbull (fl. 1780s) writing in 1785, that it was "the only work of this kind, in our language."[4]

Although Martin's *Essay* contains several passages discussing the management and treatment of enslaved people, the institution of slavery itself is only briefly examined. Unlike many overtly proslavery tracts, there is no explicit discussion of the legality of slavery and only a rudimentary justification of the slave trade based on the idea that captive Africans are prisoners of legitimate wars. These dubious but commonplace apologetics are largely missing, even though they are sometimes obliquely alluded to. Instead, there are simply, in Martin's blunt calculus, "negroes, cattle, mules, and horses." These, says Martin, "are the *nerves* of a sugar plantation." For Martin, as for many colonists and slaveholders like him, enslaved people were counted alongside the domesticated animals that likewise labored under their command. Nevertheless, Martin goes to great pains to establish himself personally as a reasonable and humane man, albeit one mindful of the bottom line. He opens his pamphlet with almost ten pages of ostensibly balanced considerations on the management of enslaved workers, including a long passage that neatly exemplifies both his humanely paternalistic persona and his more brutal underlying economic interests. "First then," he begins, "let us consider what is the right or rational method of treating negroes: For, rational Beings they are, and ought to be treated accordingly; that is, with humanity and benevolence as our fellow creatures, created by the same Almighty hand, and under the same gracious providence." Sentiments such as this would not be out of place in an abolitionist pamphlet, even if a power relation is structured in from the start. Martin then slips in a justification for slavery disguised as encouragement for benevolence: "The subordination of men to each other in society is essentially necessary to the good of the whole; and is therefore a just reason for the benevolence of superiors to their inferiors." This may not have been a particularly controversial view in 1750, either in England or the colonies, but Martin knew that there was a vast difference between the condition of an English laborer on an aristocratic estate and an African slave on a colonial plantation. He attempts to gloss this over, arguing that "the distinction between the vulgar of most free government, and the negroes of our colonies, is little more than *nominal:* for, he that is forced to hard labor in support of life, is as much *a slave to necessity,* as negroes are to their owners." This is specious reasoning, of course, but it leads him into an extended defense of slavery masquerading as a variety of Whiggish humanitarianism:

> Negroes purchased by the British planters are redeemed from absolute slavery to a degree of liberty, much greater than the subjects of absolute power in Europe: for they enjoy the privilege of written laws, by which their lives and little properties are secured; and being subject only to such correction as children are to parents, or apprentices to masters, seldom receive more punishment for felonies, than criminals do in Britain for petty larcenies: for, as it is the interest of every planter to preserve his negroes in health and strength; so, every act of cruelty is not less repugnant to the master's real profit, than it is contrary to the laws of humanity: and if a manager considers his own *case,* and his employer's interest (as he must do if an honest man) he will treat all negroes under his care with due benevolence.

The argument that it is better to be a slave than to be French was a flag-waver. It is possible that some of Martin's readers believed it, but it seems unlikely. Likewise, Martin's hint that Africans in Africa are constrained within a system of "absolute slavery" alludes to some of the justifications for the slave trade that argued it was acceptable to buy enslaved people from Africa because slavery under Europeans was somehow less tyrannical than slavery under Africans—although precisely why is seldom made clear. Martin seems to assume that his readers would have been familiar with this defense, since he develops it no further. Instead, he launches into a passage that is literally patronizing in that it speaks of enslaved people under masters as children under the subjection of parents. The key point is that this extended section of the book represents the plantation not as it actually was, but as planters wanted the world to see it. The rest of the pamphlet contains a great deal of practical information about cultivating and refining sugarcane in which enslaved people are rarely mentioned, but this is because the opening pages had already done the public-relations work. His arguments in favor of a supposedly ameliorated form of slavery cannot have been a direct response to any particular antislavery movement, at least in the earlier editions of the book. Instead, he appears to be responding to a generalized and unfocused metropolitan distaste for the business of slavery, and his final argument, while seemingly directed toward his Antiguan colleagues, might well have been aimed at a wider audience. Cruelty is not profitable, he says. This could be taken as an admonition to other Caribbean planters, but it was more likely a red herring for concerned readers in England (where the book was quickly published), assuring them that planters had the welfare of their enslaved laborers closest to both their hearts and their wallets.[5]

When it comes to the practical details about planting, manuring, weeding, and harvesting, Martin does not differ much from earlier writers like Ligon who had offered advice to planters. For example, Martin argues that good management of the soil is the business of any successful sugar planter "because upon the right culture of his soil depends absolutely the *quantity*, and in a great measure the *quality* also of his produce." Like Sloane, Martin is aware of the "dung crisis." He warns his readers of the situation in which "soils are exhausted of their fertility, by long and injudicious culture." A long chapter discussing both the necessity of appropriate manuring and the danger of poor soil management is followed by another on weeds and pests. Martin cautions that "sure disappointment will attend the planter who neglects timely weeding his young canes." This sound practical advice leads to a more assertive rejection of speculative agriculture in favor of empirically based farming. In a discussion of "that pernicious insect called the blast" (apparently the canefly, *Saccharosydne saccharivora*), Martin is at pains to reject theoretical speculations about the management of the pest:

> All the schemes hitherto proposed for curing that evil, that forerunner of the planter's ruin, are vain, impracticable speculation, or a waste of time and labor to little purpose: for as human wit penetrates no deeper than the external surface of matter; so we cannot trace the influence and effects of material bodies acting upon each other one step farther than experience, or apparent real fact is our guide. If then it be a *fact* that the blast commits most ravage in a poor land, and affects not the luxuriant sugar-canes of a rich soil; the cure of that evil is certain and obvious: *manure and cultivate* your lands so as to become rich and fruitful, and you will for ever prevent the blast.

Martin's bluff commonsense manner puts him at odds with the fashionable world of natural historians such as Griffith Hughes, who also pondered the nature of the "yellow blast" at length. Martin's approach was no doubt popular among those many planters who saw themselves as men of action and commerce, however, and like many solutions that arise from observation and experience, his cure for the blast may well have been effective, even if he could not say why. Martin was, however, at odds with the majority of planters in his thoughts on plantation diversity. In the face of a plantation system that had already effectively created a regional monoculture, Martin believes it a "matter of just concern, to see the cultivation of provisions so generally neglected in our colonies." His stated reasoning is ameliorationist. "This general neglect," he suggests, "is one great cause of our general poverty; for, while the planter feeds his negroes scantily, or with unwholesome food, how can he expect much labor and plentiful crops?" Martin's concern for the welfare of the enslaved people on his plantation does not ring entirely true since his interest is clearly economic, but the remedy is the same: diversity and crop rotation. Martin argues that "he therefore who will reap plentifully, must plant great abundance of provisions as well as sugar canes: and it is nature's œconomy so to fructify the soil by the growth of yams, plantains, and potatoes, as to yield better harvests of sugar by that very means." None of these crops in fact fix nitrogen or replenish the soil in any other obvious way. As Anna M. Foy has pointed out, the normally plain-speaking Martin "offers no description of the process by which the cultivation of yams, plantains, and potatoes might be expected to fertilize the soil in which sugarcane grows," a decorous silence that suggests he was probably advocating the use of human manure, a widespread and common practice in the period. Nevertheless, whatever the operational details may have been, there is not much evidence that his calls for diversity and rotation were widely heeded.[6]

Martin's passage on provisions and crop rotation leads immediately into another that recommends methods to ensure "the longevity of negroes." In addition to "plenty of wholesome food," he argues, "there are also other means equally necessary to strength, and the longevity of negroes well worth the planter's attention; and those are, to chuse airy, dry situations for their houses; and to observe frequently that they be kept clean, in good repair, and perfectly watertight: for, nastiness, and the inclemencies of weather generate the malignant diseases." This is no doubt true, but it is chilling in its casual commodification of enslaved people. For Martin, caring for slaves is a type of husbandry, and this is made explicit shortly after when he argues with an unnervingly unctuous tone that "then ought every planter to treat his negroes with tenderness and generosity, that they may be induced to love and obey him out of meer gratitude." This disturbing passage has prompted George Boulukos to observe that "clearly, an interest in eliciting 'gratitude' from slaves is by no means an indicator of anti-slavery, or even non-racist positions. Colonel Martin provides an excellent illustration of the self-interested nature of amelioration." Martin concludes by explicitly linking enslaved people with farmyard animals. Troublingly, his description of the care of enslaved people is almost entirely interchangeable with his description of the care of animals. "Having thus hinted the duties of a planter to his negroes," he continues, "let the next care be of his cattle, mules, and horses: for these creatures are next in degree valuable to their owner." In the extended passage on animal husbandry that follows, he writes: "Let then the good planter be kind not only to his fellow creatures, but merciful to his beasts even for the sake of the owner; giving them plenty and variety of wholesome food, clear water, cool shade, and a clean bed; bleeding them after a long course of hard labour, currying their hides from filth and ticks, affording them salt, and other physick, such as experience prescribes; protecting them from the flaying rope lashes of a cruel driver (who needs no other instrument than a goad) proportioning their labor to their strength; and by every art rendering their work as easy as possible." This passage underscores that any concern Martin had for animals was primarily motivated by economic benefit. Likewise, anyone tempted to think that his seemingly ameliorative description of the treatment of enslaved people was motivated by anything other than profit should be undeceived at this point. For the third time in a short book, he explicitly likens treatment of people with treatment of animals, and in both cases makes clear that his prescription is primarily "for the sake of the owner." The language that he uses to speak of animals is not far from that he uses to describe the enslaved people on the plantation, and while on the surface he appears to be urging humane treatment, his criticism of excessive labor and the cruel use of the whip simply reminds us that such cruelty was the norm rather than the exception, both to people and to animals. Whether openly cruel or ostensibly humane, Martin's plantation is above all an industrial unit premised on exploitation. Fearing "infinite expense of time, both of cattle and negroes" and echoing Newtonian conceptions of the cosmos as a vast clockwork, Martin sees the plantation as an unstoppable machine that must run its course no matter what. "A plantation," he tells us, "ought to be considered as a well constructed machine, compounded of various wheels, turning different ways, and yet all contributing to the great end proposed." That "great end" is simply profit, and there are many ways in which perturbations of the machine can diminish returns: "if any one part runs too fast or too slow in proportion to the rest, the main purpose is defeated." As to those parts, the "various wheels," Martin clearly means livestock and enslaved people. "It is in vain to plead in excuse the want of hands or cattle," he admonishes would-be planters, "because these wants must either be supplied, or the planter must contract his views, and proportion them to his abilities." In other words, the plantation must proceed under its own logic and toward its own ends, it must turn under the labor of "Negroes, cattle, mules, and horses," and it must lead inevitably either to profit, under the wise management of a planter, or else "the attempt to do more than can be attained, will lead into perpetual disorder, and conclude in poverty." Martin's mechanistic plantation demands an appropriate managerial response from the planter rather than allowing him the freedom to direct as he chooses. Martin is in general an uncomplicated commonsense thinker, but here he pushes common sense to its philosophical limits, presenting a reductionist and deterministic model of a complex system that echoes a then fashionable conception of the nature of the universe itself. In Martin's view, the plantation inexorably follows the laws of nature, and its cogs and wheels, its domestic animals and enslaved Africans, can no more be replaced than can the stars and planets of the cosmos itself. The most influential plantation manual was thus in agreement with the most prominent natural histories of the day. Slavery was an intrinsic feature of the Caribbean natural environment that needed to be managed rather than a contingent social institution that might be abolished.[7]

MANAGEMENT VERSIFIED: JAMES GRAINGER'S THE SUGAR-CANE

An agricultural manual might seem an unlikely source of literary inspiration, but Martin nonetheless inspired at least two authors, both Scots.[8] The ameliorationist novel *Julia de Roubigné,* published in 1777 by Henry Mackenzie (1745–1831), echoes Samuel Martin's description of the plantation as a machine. The novel's protagonist, seeking to make his fortune on a plantation in Martinique, observes: "I have often been tempted to doubt whether there is not an error in the whole plan of negro servitude, and whether whites, or creoles born in the West-Indies, or perhaps cattle, after the manner of European husbandry, would not do the business better and cheaper than the slaves do. The money which the latter cost at first, the sickness (often owing to despondency of mind) to which they are liable after their arrival, and the proportion that die in consequence of it, make the machine, if it may be so called, of a plantation extremely expensive in its operations." Mackenzie's comparison of enslaved people to domestic animals, his insistence that slaves are essentially economic units in a plantation machine, and his repeated argument that a humane plantation is a more profitable one, indicate that Martin's *Essay* was an important source for the reformed plantation system he proposed in *Julia de Roubigné.* Contemporary readers may have recognized this immediately, since Mackenzie was not the only literary author to turn to Martin. In 1764, James Grainger (c. 1721–c. 1766) eulogized a certain "M" in his long georgic poem *The Sugar-Cane:*

> O M * * * ! thou, whose polish'd mind contains
> Each science useful to thy native isle!
> Philosopher, without the hermit's spleen!
> Polite, yet learned; and, tho' solid, gay!
> Critic, whose head each beauty, fond, admires;
> Whose heart each error flings in friendly shade!
> Planter, whose youth sage cultivation taught
> Each secret lesson of her sylvan school.

To avoid any doubt about the identity of "M," Grainger prefaced his poem with the observation "I have often been astonished, that so little has been published on the cultivation of the Sugar-Cane, while the press has groaned under folios on every other branch of rural oeconomy. . . . An Essay, by Colonel Martyn of Antigua is the only piece on plantership I have seen deserving a perusal. That gentleman's pamphlet is, indeed, an excellent performance; and to it I own myself indebted."[9]

The Sugar-Cane, a poem of 2,562 lines divided into four books, has been described as an example of "imperial georgic." Grainger himself called it "West-India georgic." On the face of it, this makes it an unlikely candidate for a practical plantation management manual. In fact, the poem has always been viewed as akin to that tradition: the slavery apologist Gordon Turnbull, for instance, immediately qualified his 1785 observation that Martin's *Essay* is "the only work of this kind" by adding, "unless *The Sugar-Cane,* a poem, by Dr Grainger, may be stiled a practical treatise on the subject." The poem's credentials as a "practical treatise" arise variously from its georgic form, its debt to Martin's *Essay,* and its author's claim to experience. Grainger was a doctor and would-be writer who moved in circles that included Samuel Johnson (1709–84), Oliver Goldsmith (1728–74), and Tobias Smollett (1721–71), although the connections did him little good. He left London for four years in St. Kitts, where he set up a medical practice, married a local woman, bought slaves, and employed himself with botany and versification. By 1763, *The Sugar-Cane* was complete, and Grainger returned to London to find a publisher. Famously, he read the poem to a group of literary luminaries at the house of Sir Joshua Reynolds (1723–92). According to James Boswell (1740–95), "all the assembled wits burst into a laugh" when Grainger reached the line "Now, Muse, let's sing of rats." Worse still, "what increased the ridicule was, that one of the company, who slily overlooked the reader, perceived that the word had been originally MICE, and had been altered to RATS, as more dignified." This delicious and often-told story perhaps reveals the low opinion the metropolitan literary elite had of colonial literary pretensions but, as John Chalker shows, Grainger's attempts to poeticize rats in the mock-heroic mode were in fact carefully controlled and underlying *The Sugar-Cane* is "the epic theme of man's struggle against Nature, and sometimes this struggle manifests itself as one against rats." Beccie Puneet Randhawa draws attention to the fact that Grainger traveled halfway around the world to read out his poem: "Clearly," she notes, "Grainger craved the cultural approbation of London's writing establishment as the ultimate validation of his poetry." The poem was published shortly thereafter, and Grainger returned to St. Kitts, where he set himself up in a fine house with a spacious library, no doubt intending further literary work. This, however, was not to be; he soon died of a fever, probably in December 1766.[10]

The Sugar-Cane was plainly built on Grainger's personal experience as a professional doctor and amateur botanist in St. Kitts, but it also emerges from the long tradition of instructional agricultural verse that began in the classical period and had been undergoing something of a renaissance in eighteenth-century Britain with such titles as John Philips's (1676–1709) *Cyder* (1708) and John Dyer's (1699–1757) *The Fleece* (1757). The form had its origins with Virgil's *Georgics,* a work that Grainger knew well and on which he modeled his poem. As John Gilmore has shown, "many aspects of *The Sugar-Cane* are based on conscious imitation of Virgil's *Georgics.* These begin with the structure of the poem in four books, which begin with the soil itself and progress to the slaves and their treatment." Gilmore presents a line-by-line commentary that shows Grainger often adheres meticulously to Virgil, sometimes approaching direct translation, with the substitution of West Indian place and species names. Grainger's debt to Virgil would have been readily apparent to educated mid-eighteenth-century readers but, as a form, georgic verse had always served two audiences. On the one hand, it appeared to offer practical advice on agricultural topics, ostensibly appealing to landowners with a taste for elegant verse. At the same time, georgic represented an idealized image of rural life to urban readers who probably had little practical interest in agriculture. Thus, despite Turnbull's hesitant inclusion of *The Sugar-Cane* as a "practical treatise," it is doubtful that many planters would have referred to the poem as a manual for their daily business. As Markman Ellis has argued, "The apparent addressee of the 'West-India georgic' is the planter in the colonies, but the implied audience is the metropolitan political and literary elite." Although the poem "sold well amongst the slave-owning planters of the Caribbean," this was unlikely for its practical use but rather because the poem "afforded evidence of their own cultural standing in the metropolis." The snobbish values of Boswell and his crowd were exactly those to which the planters aspired. Indeed, as Anna M. Foy has shown in an important reading of the poem's manuscript, Grainger moderated his initial high praise for the planters better to suit a metropolitan audience—although whether this was colonial strategy, personal aggrandizement, or both is unclear.[11]

The Sugar-Cane consciously covers much of the same ground as Martin's *Essay* and, inevitably, has much to say about weeds, soil, and dung. Grainger's defensiveness when introducing such earthy topics is palpable, but he rises to the challenge, for example when he asks:

Of composts shall the Muse descend to sing,
Nor soil her heavenly plumes? The sacred Muse
Nought sordid deems, but what is base; nought fair
Unless true Virtue stamp it with her seal.
Then, Planter, wouldst thou double thine estate;
Never, ah never, be asham'd to tread
Thy dung-heaps, where the refuse of thy mills,
With all the ashes, all thy coppers yield,
With weeds, mould, dung, and stale, a compost form,
Of force to fertilize the poorest soil.

Grainger's rallying cry to planters to walk their dung heaps with pride may appear faintly ludicrous, an example of what Karen O'Brien has called the poem's "ethical as well as tonal bathos." It may well have amused Boswell and his crowd. As Gilmore points out, however, it directly imitates Virgil and, as Christopher F. Loar notes, it offers a "vision of sustainable agriculture in the Caribbean." Indeed, despite its apparent defensiveness, it articulates a growing sense of colonial purpose, identity, and pride that within a decade would express itself in rebellion and revolution in the continental colonies of North America, even if Caribbean planters chose to remain British. It invites planters to celebrate their own abilities to colonize inhospitable tropical environments and to bring them into cultivation. As such, *The Sugar-Cane* can also be seen as the most substantial of a small group of poems that sought to reconcile Old World poetic tradition with the New through the deployment of a vocabulary of Caribbean fauna, flora, and agriculture. Located both temporally and geographically between the 1754 *Barbados: A Poem,* written by Nathaniel Weekes's (fl. 1752–65) and the 1767 *A General Description of the West-Indian Islands* by John Singleton (fl. 1767), *The Sugar-Cane,* argues Emily Senior, offers "an agricultural metaphor to describe the great potential for aesthetic and natural philosophical development presented by the Caribbean islands." We may also note that, by yoking together Martin's treatise on plantation management with Virgil's poeticized account of Italian agriculture, Grainger is implicitly making the claim that the Caribbean plantation is a natural extension of the classical farm, which relied on enslaved laborers.[12]

As well as Martin's *Essay,* Grainger also draws extensively on the work of naturalists, including Hans Sloane, whom he cites in his notes, and Griffiths Hughes, who is not explicitly mentioned but who is a clear influence; Grainger emulates Hughes in his use of colonial vernacular names when describing tropical species and his frontispiece, a botanical illustration of the sugar-cane,

Figure 2. Frontispiece illustration of sugarcane. From James Grainger, *The Sugar-Cane: A Poem*, 1764. (© The British Library Board.)

Figure 3. Indian corn and sugarcane. From Griffith Hughes, *The Natural History of Barbados*, 1750. (© The British Library Board.)

is an exact copy of the sugarcane illustrated in Hughes's *Natural History of Barbados* (see figures 2 and 3). It is not easy, however, to tease out Grainger's underlying philosophy which, like his poem, seems eclectic and various. At times his verse is rhapsodic in its praise of nature, as for instance when he courts potential colonists by dangling in front of them the promise of riches, power, and tropical beauty. To young men back home in Britain he asks:

> Doth the love of nature charm;
> Its mighty love your chief attention claim?
> Leave Europe; there, through all her coyest ways,
> Her secret mazes, nature is pursued:
> But here, with savage loneliness, she reigns
> On yonder peak, whence giddy fancy looks,
> Affrighted, on the labouring main below.

This passage in part accords with Shaun Irlam's conception of the poem as "a conspicuous attempt to recruit metropolitan sanction. . . . Grainger makes an earnest appeal for the approval of the metropolitan literary elite; the poem becomes a poignant postcard from abroad, effectively writing its own version of 'Wish You Were Here.' " It does not sit quite so comfortably, however, with Irlam's contention that the poem seeks to "domesticate the foreign, Caribbean terrain in terms of familiar social, literary, and agricultural codes produced within the already constructed georgic discourse of the English landscape." Instead, the idea of the "savage loneliness" of the Caribbean is invoked in opposition to the neatly managed landscapes of Europe. This appeals to then newly fashionable ideas about the sublime and picturesque in landscape that would reach their fullest expression with the poets of the Romantic era. Nevertheless, in Grainger's analysis even the wildest natural landscape is ultimately to be tamed and exploited. The mountains are full of "sulphurs, ores, what earths and stones abound." These are the appropriate subject of "philosophy," he argues, since " 'For naught is useless made.' " Although night and day "shall hear my constant vows / To Nature," ultimately Grainger's ambition (in this section of the poem, at any rate) is to discover in nature something "of use to mortal man."[13]

Here and elsewhere, Grainger's position is shifting and ambiguous, as he negotiates between his intention to represent the physical reality of a tropical plantation and his desire to phrase himself in a neoclassical poetic diction that was itself being modified by contemporary notions of the sublime. His extended discussion of the insect pest known as "the blast" is a case in point. Martin had consulted his long experience and argued that it could best be countered by manuring the soil well. Grainger is less sanguine, presenting the attempt as an unwinnable battle with an implacable invader:

> And pity the poor planter; when the blast,
> Fell plague of Heaven! perdition of the isles!
> Attacks his waving gold. Tho' well-manur'd;
> A richness tho' thy fields from nature boast;
> Though seasons pour; this pestilence invades.

As Monique Allewaert has demonstrated, although recent critics have shown most interest in book 4 of *The Sugar-Cane,* with its depiction of slavery, Grainger considered the accounts of natural phenomena in book 2 to be his work's centerpiece. This book, Allewaert notes, "oscillates between insectophilia and insecticide" in its depiction of both harmful and benign insects, while representing them through complex forms of metropolitan and colonial personification; the former "casts personification's animating power as an affective operation, and it uses this operation to join the diversity that it collocates into a single system," while the latter "casts personification's animating power as a disaffecting operation of the small and the particulate and, instead of working toward connection, it tends toward division." Grainger's account of the blast leans toward the latter, representing the massed insects as a conquering army of "bugs confederate, in destructive league" that, despite their confederation, are most dangerous because they are divided into numerous units that must be dealt with individually by deploying the plantation's defending army of enslaved laborers. Planters hopeful of success must

> Command their slaves each tainted blade to pick
> With care, and burn them in vindictive flames.
> Labour immense! and yet, if small the pest;
> If numerous, if industrious be thy gang;
> At length, thou may'st the victory obtain.
> But, if the living taint be far diffus'd,
> Bootless this toil.

He concludes, somewhat hopelessly, that burning the entire crop may be the only remedy: "Far better, thus, a mighty loss sustain, / Which happier years and prudence may retrieve." Although the pest is personified as a conquering horde, his argument that it is simply a "Fell plague of Heaven" points to a conventional hierarchical view of the environment as simply a location for the unfolding of God's plan, and in this he appears to reject both Ligon's proto-ecologies and Martin's empirical common sense. But this fatalism is not a stance he adopts elsewhere in the poem or its apparatus. His extensive notes often read like extracts from a conventional natural history, making use of both printed sources and his own observations on St. Kitts. The method is both empirical and systematic, suggesting that, had he lived, Grainger may have produced a more thorough natural history of the island to add to his four medical works, including an *Essay on the More Common West-India Diseases,* published in 1764, which offered both botanical remedies and plantation management advice. In his notes to *The Sugar-Cane,* Grainger even undercuts his own poetic persona at times, not least in the footnote to his line on the blast where, though decrying manuring as an ineffectual cure in verse, in prose he reverts to Martin's authority and concedes that "it must, however, be confessed, that the blast is less frequent in lands naturally rich, or such as are made so by well-rotted manure." While the blast section of the poem reads like an ecological conflict between cane, canefly, and planter, the notes themselves speak of a conflict between Grainger the doctor and botanist and Grainger the poet.[14]

Whether one thinks of it as inconsistency, as posturing, or as a pleasing diversity of tone and register in a miscellaneous poem, Grainger's ability to speak from multiple and sometimes contradictory positions is a prominent feature of *The Sugar-Cane.* Britt Rusert has argued that in this respect the poem represents "the experimental plantation," a formulation "that includes scientific, medical, and agricultural experimentation on the plantation, as well as the profligate exploitation of enslaved laborers within 'experimental' regimes of plantation surveillance and punishment." As well as depicting this regime, Rusert contends, Grainger also advances it: "The experimental plantation also includes literary experiments produced on the plantation, including Grainger's poem." Rusert sees the notes in particular as Grainger's opportunity to take "his expertise as a physician seriously as he crams all kinds of botanical, medical, and scientific knowledge about the Caribbean into sprawling footnotes that threaten to overwhelm the body of the poem itself." Rusert's analysis helps us understand the poem's eclectic nature as well its internal conflict between physician and poet, but it also offers an insight into why the content is often incongruous and the tone discordant—and nowhere is the incongruity of Grainger's experimental plantation more evident than in his discussion of the enslaved people of St. Kitts. Grainger announced in the fourth line of the first book that an important part of the poem would be discussion of how planters should treat "Afric's sable progeny." This discussion comes in the final book, but before that enslaved people appear only glancingly, and these appearances are neither convincing nor naturalistic. For instance, apparently echoing Ligon, Grainger contrasts the plantation overseer with "some delegated chief" of a monarch who "rushes to the war." The battle lines are the furrows of the cane fields, but the delegated chief's orders are simply "let not thy Blacks irregularly hoe." This epic simile is quoted by John Chalker as "typical" of Grainger's "staple style" of "considerable elevation," consistent with the poem's epic theme of "the struggle of man against the afflictions of nature," but the effect is in fact unintentionally bathetic and, in this, is characteristic of the way in which Grainger's subject and style frequently come into conflict.[15]

Even less credible is Grainger's depiction, a few lines later, of the "Negroe-train" that at the end of the day is allowed to "disperse, all-jocund, o'er the long-hoed land." This image of enslaved laborers joyfully dispersing after an honest day's work elides the fact that the labor had been dishonestly extorted from captive people who had been violently coerced. Grainger compounds this in the third book when he represents enslaved Africans as eager workers who "pant to wield the bill" with "willing ardour" and who cheerfully sing while they work—although he gives no indication of knowing what their songs might be about. Echoing Martin's ameliorative stance, Grainger uses the image to argue for more lenient discipline. Harking back to the notion of the overseer as "some delegated chief," he cautions, " 'Tis malice now, 'tis wantonness of power / To lash the laughing, labouring, singing throng." Grainger articulates a more comprehensive ameliorative program later in the poem, but here one is startled by the words "laughing" and "singing." Of course, enslaved people working the fields talked, sang, and no doubt sometimes laughed, but Grainger's joyful scene is unconvincing at best. A century later, following his own escape from enslavement, Frederick Douglass (c. 1817–95) remarked, "I have often been utterly astonished, since I came to the north, to find persons who could speak of the singing, among slaves, as evidence of their contentment and happiness. It is impossible to conceive of a greater mistake. Slaves sing most when they are most unhappy." Whether Grainger's mistake, in Douglass's terms, is genuine or intended, it is unlikely that even the most cynical apologist for slavery would genuinely have been persuaded by these rose-tinted representations of plantation life, whatever they may have said in public. Either way, these scenes lack detail and focus as well as plausibility. For Grainger, as for many plantation management writers and natural historians of the Caribbean before him, enslaved people are for the most part simply picturesque figures in a sublime landscape whose genuine human experience can be safely ignored or rendered innocuous by poetic diction.[16]

In the final book of the poem, which is euphemistically dedicated to the "Genius of Africk" (the spirit of Africa), Grainger develops an extended discussion of the purchase, characters, and treatment of enslaved people on the plantation. Again, the debt to Martin is clear when Grainger urges planters to "let humanity prevail," and, like Martin, he links humane treatment to increased productivity. He also strongly asserts that "howe'er insensate some may deem their slaves," Africans are fully human: "The Ethiop knows, / The Ethiop feels, when treated like a man." In a passage that has attracted much critical attention, he even appears to bring the whole institution of slavery into question. He laments:

Oh, did the tender muse possess the power,
Which monarchs have, and monarchs oft abuse:
'Twould be the fond ambition of her soul,
To quell tyrannic sway; knock off the chains
Of heart-debasing slavery; give to man,
Of every colour and of every clime,
Freedom, which stamps him image of his God.
Then laws, Oppression's scourge, fair Virtue's prop,
Offspring of Wisdom! should impartial reign,
To knit the whole in well-accorded strife:
Servants, not slaves; of choice, and not compell'd;
The Blacks should cultivate the Cane-land isles.

Responses to this passage have been varied. Jim Egan draws attention to its emphasis on climate as a marker of national and racial difference, an example of Grainger's repeated insistence on climate as the foundation of identity throughout the poem. Steven W. Thomas suggests the passage raises more questions about colonial administration than it answers and notes the passage's "abrupt shift from a political problem to a medical one" at the end, when a new paragraph begins talking about worm infestation—part of a move by which Grainger "racialized both servitude and disease as black problems." In response to the apparent antislavery sentiment of this passage, Keith Sandiford has argued that "the moral evil of slavery itself is acknowledged only within the unthreatening context of the visionary imagination," while Markman Ellis has noted that it is only the muse, "a mere woman, and a poetic fancy" who is entertaining ideas of ending slavery. In Grainger's verse, Ellis argues, "not only is the emancipation proposition fanciful . . . but also it is a fantasy induced by literary convention." Cristobal Silva concurs that the "abolitionist sentiment in the poem is fleeting, and ultimately rooted in a form of poetic failure. That is, despite any desire to the contrary, Grainger's muse is ultimately powerless to 'quell tyrannic sway'—either in the poem, or in the West Indian world it describes."[17]

Silva acknowledges that "excerpting these few lines gives a distorted picture of *The Sugar-Cane*'s overall stance toward slavery" but argues that "the passage itself is crucial to understanding that Grainger used the poem as a testing ground for his ambivalence about the practice." This is a forgiving reading of the poem. Grainger's miscellaneous style does allow him to pose temporarily as an early version of an abolitionist, which might appear as ambivalence, but the volume of proslavery material in the poem strongly indicates that he is quite the opposite. Grainger's *Sugar-Cane* is essentially Martin's *Essay upon Plantership* versified, with the ameliorationist sentiment little more than a smokescreen for protecting the profits and preserving the system of West Indian slavery. The muse does not possess the power to end slavery nor, by extension, do the poet or the planter class he represents. Even if monarchs do theoretically wield that power, the poem makes it clear that this power is rarely exercised—indeed, is often abused. Not only is the system cast as natural and permanent but, once the poetic diction has been peeled away, Grainger's writing appears more openly complicit with the brutal realities of slavery and the slave market than Martin's. Grainger's humanitarian rhetoric sits uncomfortably with the poem's lengthy passages offering advice on buying enslaved people in the market, which even include an overt encouragement to inflict violence:

> When first your Blacks are novel to the hoe;
> Study their humours: Some, soft-soothing words;
> Some, presents; and some, menaces subdue;
> And some I've known, so stubborn is their kind,
> Whom blows, alas! could win alone to toil.

With the "alas," Grainger again postures as a benevolent planter regretting the use of force, but it does not ring true after the studied bribes and menaces, even though, in the following line, he again changes tack and exclaims "Yet, planter, let humanity prevail." The plea sounds hollow and Grainger appears calculating and sadistic, happy to use violence, perhaps as a last resort, but happy all the same. In this formula, the end, the cultivation of the sugarcane and the profits it supplies, justifies the means, while the cane and those forced to cultivate it become bound together in what Martin had described as a "well constructed machine, compounded of various wheels, turning different ways, and yet all contributing to the great end proposed." Force might be spared, but only if the wheels keep turning.[18]

Grainger was clearly a useful doctor and a competent botanist. His descriptions of the flora of the island and of diseases among the enslaved population of St. Kitts remain of interest to scientists today. The poem is one of the most substantial literary productions to emerge from the plantations of the British Caribbean, and it displays its debt both to natural histories of the island and to plantation management literature. One element of its importance is its attempt to translate the rough world of colonial sugar production to the polite world of the literary salon—although it can equally be viewed as an attempt to bring the salon to the plantation. More significantly, we can read it as part of an attempt to recover the georgic poem to describe new labor relations in an age both of colonial slavery and of metropolitan agrarian reform. Grainger's *Sugar-Cane* represents the opinions of a confident and largely unopposed plantocracy that had yet to come under sustained attack from abolitionist authors or activists. It articulates with impressive, if disturbing, clarity the aesthetic and scientific vision of an empire of introduced species imposed by European settlers on a tropical landscape and maintained by imported labor, imported agricultural techniques, and imported cultural norms. If an unnatural empire of the type Grainger describes needs to be performed as well as planted, then arguably *The Sugar-Cane* is its bravura performance.

RACIALIZING THE PLANTATION: EDWARD LONG

By the early 1770s, there were increasing indications that slavery and the slave trade were no longer universally tolerated in Great Britain and some of its American colonies, not even as unpleasant but necessary evils. Quakers, as early as the 1750s in Philadelphia and in 1761 in London, had issued strong condemnations of slavery and, by 1774, the Methodist leader John Wesley (1703–91) had published his *Thoughts upon Slavery* condemning both slavery and the slave trade. In London, the legal campaigner Granville Sharp (1735–1813) had challenged the legality of slavery on English soil, culminating in Lord Mansfield's ruling in the 1771–72 case of James Somerset, an enslaved African in London who was facing a forced passage to the West Indies. He absconded, but was brought before the court. Lord Chief Justice Mansfield's highly publicized decision in the case, made in June 1772, freed Somerset but did not technically make slavery in England illegal. Instead, Mansfield gave the legal opinion that slavery had never been positively introduced into English law, and that therefore no slaveholder had the authority to compel their slave to leave the country against their will. In practical terms, this meant that any enslaved person in England could walk away from their servitude without fear of legitimate recapture, which made enslavement almost impossible to enforce in England by any legal means. This was immediately, if erroneously, interpreted to mean that slavery had been outlawed in England, but while the change to the law was in fact not quite so clear-cut, what was apparent to all was that a significant segment of the English public was no longer comfortable with accepting slavery at home, whatever they might have tolerated overseas.[19]

It was in this context that the lawyer and former colonial administrator Edward Long (1734–1813) published his three-volume *History of Jamaica* in London in 1774. Long's family had had interests in Jamaica, both as planters and as administrators, since the island had been captured from the Spanish in the 1650s. Long himself, although born and educated in England, lived in Jamaica for twelve years between 1757 and 1769. He was elected to the Jamaican assembly in 1761 and became its speaker in 1768. Ill health forced him to return to England the following year, and there he set about writing the *History of Jamaica,* for which he is now principally remembered. The book is notorious today for its contribution to the development of scientific racism but the articulation of these views, which are discussed below, does not appear to have been Long's primary purpose in writing the book. His intention, which he sets out succinctly in the introduction, was to give "a competent information of the establishments civil and military, and state, of Jamaica, its productions, and commerce; to speak compendiously of its agriculture; to give some account of the climate, soil, rivers, and mineral waters; with a summary description of its dependencies, counties, towns, villages, and hamlets, and the most remarkable natural curiosities hitherto discovered in it." Although Long does not claim to be a specialist botanist or zoologist, the proposed topics nevertheless encompass those found in both natural and political histories, and this includes ethnography. Long also intends, the introduction tells us, "to display an impartial character of its inhabitants of all complexions, with some strictures on the Negroe slaves in particular." Long's decision to order Jamaica's inhabitants by "complexion" reveals the depth of his racial thinking, even though he insists that *all* complexions will come under scrutiny, but we should also note that his ethnography comes late in the proposed running order, and that the "strictures on the Negroe slaves" do not appear until toward the end of the second volume, although they do occupy approximately 180 pages of the almost 1,000 total pages of the book. Before he reaches that section, and for much of the following volume, Long's interest is primarily in Jamaican political history as well as the island's economy and the environmental features and resources that underpinned it. Indeed, contemporary reviews suggest that most of Long's readers turned to the book for its political and natural history rather than for Long's extreme views on race. William Donaldson's lengthy and sympathetic assessment in the *Monthly Review,* for example, takes up seventeen pages spread over two separate issues of the journal and barely mentions Long's "strictures on the Negroe slaves" other than to "applaud his philanthropy" when Long recommends a supposedly more humane method of buying and selling enslaved people. Otherwise, the reviewer concludes, of all the topics Long covers, "Cultivation and commerce are his fondest concern."[20]

The book does indeed cover a wide range of topics. Volume 1 is almost entirely concerned with the political, commercial, and economic history of Jamaica as well as its current political and economic circumstances. In volume 2, Long makes extended descriptive circuits of each of Jamaica's three counties, Middlesex, Surrey, and Cornwall, apparently based on genuine walking tours around the counties' parishes, before concluding with a general ethnography of the island containing the notorious "strictures on the Negroe slaves." The final volume conforms more closely to the traditional structure of natural history; Long offers a hundred pages of meteorology before setting out a "Synopsis of Vegetable and Other Productions of the Island." While the headings are quite distinct, the content is more eclectic. Throughout the book, Long combines observations on the present state of the island with those on its history and repeatedly conflates Jamaica's environment with its economy. He draws on official and unofficial published sources as well as his own personal observation. The effect is compendious, but Long's personality is never submerged and his racist views surface not merely in the "strictures" but almost every time in the book he mentions African people.

The county circuits in volume 2 display Long at his most various and at his most personal and reveal an inner conflict between his admiration of the picturesque and his instinct for economic exploitation. His descriptions of Jamaica's physical and biotic environment, which intersperse observations on its local history and economy, are, however, more akin to nature writing or tourist narrative than to scientific natural history. To take one example from many, in his sixteen-page description of St. Andrew Parish, which lies immediately to the north of Kingston, he sets out on the "extremely romantic" road into the Blue Mountains. As he ascends, he looks down from a great height across "a most luxuriant and extensive landscape" consisting of "cane-fields of the liveliest verdure, pastures, and little villas intermixed; the towns and ports of Kingston and Port Royal; the shipping scattered in different groups; the forts; . . . and, beyond these, a plain of ocean extending to the Southern hemisphere." These components of a prosperous colonial landscape leading to an ocean that stretches enticingly southward to other commercial opportunities "form all together a very pleasing combination" in the eyes of Long the commercial historian and colonial administrator. His tour then leads him upward into the mountains with its picturesque landscapes and changeable weather. Anticipating some of the most famous images and poetic sequences of European Romanticism, Long witnesses "another scene, not less magnificent, though more awful" when he ascends above the clouds and sees them spread below him as "a spacious sea" in which a storm is brewing. "We hear the majestic thunder rolling at our feet, and reverberated by a thousand echoes among the hills," he records in terms that recall the poetic sublime more than the register of scientific observation, before concluding that "one seems to have been transported by some magic vehicle into a foreign country." Turning his attention to the birdlife in the hills, he notes that "the ring-tailed pigeons frequent them in great numbers: they are seen constantly on the wing, and generally darting along the fogs, which it is imagined they involve themselves in, the better to conceal their flight." Long's engaging description of the ring-tailed pigeon (*Patagioenas caribaea*), which is endemic to Jamaica, is both poeticized and speculative, attempting to imagine the birds' reasoning and motivation. He then lists the birds, fish, reptiles, and insects he encounters in terms that are subjective and impressionistic; "a large, black and yellow-striped humble bee; a fly of the cantherides kind; red and stinging ants; wasps; a beautiful, long forked tail butterfly, of a copperish and green hue. Of plants are observed a prodigious variety of ferns, and a still greater of mosses; black and bill-berry bushes in abundance, large and flourishing."[21]

Aestheticized, almost rapturous descriptions such as these are repeated for each of the three counties. One of the most poetic is his description of the "two remarkable cascades" in St. Ann Parish: the "Roaring River Cascade," today better known as the Dunn's River Falls, and the rapids on the White River. The travertine terraces of Dunn's River, in Long's view, bear "no ill resemblance to a magnificent flight of steps in rustic work, leading up to the enchanted palace of some puissant giant of romance," such that "description fails in attempting to convey any competent sense of its several beauties." The White River Falls, by contrast, are sublime rather than beautiful. "The roaring of the flood," Long effuses, "the tumultuous violence of the torrent, tumbling headlong with restless fury; and the gloom of the over-hanging wood" contrast with "the soft serenity of the sky" and "the flight of birds skimming over the lofty summit of the mountain" to form a scene "beyond the power of painting to express." Words failing him, Long inserts an engraving and a long quotation from the "Winter" (1726) of James Thomson's (1700–1748) *Seasons.* But the picturesque is rapidly overtaken by a brutal reversion to the reality of

Jamaica's role in the imperial economy as Long reaches the bottom line. The tour ends, as do all his tours, with a summary "State of the Parish" that enumerates in a table the parish's "Negroes, Cattle, Sugar-plantations, and Hogsheads," statistics Long uses "to shew the vast benefits arising to the island from a more extensive colonization of its interior wastes." As Long reverts to the mentality of the colonial administrator, unconquered nature is no longer rhapsodized as sublime landscape but is now merely derided as unproductive wasteland.[22]

Long's county and parish tours may present the observations of a tourist rather than the taxonomic precision of a trained naturalist, but they nonetheless reveal him to be an author with a keen eye for detail, a genuine appreciation of wildlife, landforms, and meteorology, and sufficient knowledge of recent science to be under no illusions both about his own limitations and the relative paucity of research into Jamaican flora and fauna. In a three-page digression in his account of the Blue Mountains, he laments that "no person, I believe, has hitherto visited them with the professed design of examining all their natural productions." Even Hans Sloane was unable to reach them and was forced to depend on the accounts of others, "hence the many inaccuracies in his work." Long calls for a qualified, properly funded person to do the work but is careful not to pitch this as pure science; the proposed work would give "the public that satisfactory knowledge of soils, climates, and productions, that, while it gratifies the *literati,* may also tend to improve and people this country." The proposal is backed up by a twin appeal to natural theology and colonial commercial opportunity combined with a swipe at "the despicable tribe of insect-hunters, and collectors of gimcracks." These amateur collectors, Long argues, have brought the science of natural history into "contempt and ridicule," but we should "be cautious to separate from this dross all those, whose labours conduce to the most useful purposes of life; who not only disclose to us the wonderful mechanism of the creation, and the wisdom of the Deity; but exemplify his unbounded benevolence to man, while they instruct us in the means by which our health may be preserved, our life prolonged, our agriculture improved, manufactures enlarged and multiplied, commerce and trade extended, and the public enriched." For Long, the Caribbean environment is a resource supplied by God for the benefit of the British people and economy, and natural history is a tool in its exploitation.[23]

Although historians and critics have tended to read them in isolation, Long's notorious comments on race must be understood in the context of this wider vision of a colonial world in which both land and people are naturally fitted to meet the requirements of an expanding commercial empire. The long chapter at the end of volume 2 simply titled "Negroes," in which Long articulates some of the most clearly expressed and carefully formulated racist views of the century, is not separate from but is an integral part of his natural history of Jamaica, designed to show that the enslavement of Africans is not merely justifiable on the grounds of national interest but is natural and unavoidable. This is articulated from the outset. In his introduction, Long responds to Lord Mansfield's ruling that "the state of slavery is of such a nature that it is incapable of being introduced on any reasons, moral or political, but only by positive law." Long does not dispute this directly, instead confronting "the enemies of the West-India islands" by reminding them that slavery "was tolerated by both the *Romans* and *Athenians,*" who were "jealous of their own liberty," before apparently accepting Mansfield's interpretation of English law. "What I have said does not imply," he argues, "that a system of servitude ought to be introduced into any free country; but only mean to shew, that it may be permitted with least disadvantage, both to the master and vassal, in those parts of the world, where it happens to be *inevitably* necessary." In other words, while Long is prepared to accept the Mansfield ruling for England, he believes that it strengthens the view that slavery in other parts of the world is natural or, as he phrases it, *inevitable.* Since enslaved Africans are not naturally found in Jamaica, Long has to prove, first, that enslavement is natural in Africa and second, that African enslavement is or can be naturalized to Jamaica. Long's attempts to meet this aim follows a pattern similar to his engagements with the Jamaican environment as he brings to bear the scientific knowledge of a well-read but nonexpert layman to support his own personal observations and prejudices.[24]

First, Long seeks to prove that Africans are not of the same biological species as Europeans. He builds on some shorter comments in an earlier chapter, "Freed Blacks and Mulattos," where he offers the observation that, in his own personal experience, "mulatto" women, that is, women with mixed African and European parentage, "either proving barren, or their offspring, if they have any, not attaining to maturity . . . tends to establish an opinion, which several have entertained, that the White and the Negroe had not one common origin." His argument is tautological, starting with his own personal observations, finding that the theories of others support his observations, and concluding by reaffirming his own opinion with the statement "For my own part, I think there are extremely potent reasons for believing, that the White and the Negroe are two distinct species." These "potent reasons" are expanded on in considerable depth in the chapter "Negroes," which extends for 155 pages and contains many of the book's most notorious statements of racial hatred. It begins by emphasizing the physical characters by which he thought African people more closely resembled nonhuman animals than European people. These include African hair, which he called "a covering of wool, like the bestial fleece," and what he describes as the "bestial or fetid smell" of African people. He denigrates or denies every intellectual, moral, or civic capability that African people might possess, arguing, "In general, they are void of genius, and seem almost incapable of making any progress in civility or science. They have no plan or system of morality among them. Their barbarity to their children debases their nature even below that of brutes." He concludes his opening salvo by revealing his own confirmation bias. When he argues that "they are represented by all authors as the vilest of the human kind, to which they have little more pretension of resemblance than what arises from their exterior form," it is clear that he has not consulted "all authors" but merely those that confirm his prejudices. In the following pages, he constructs an elaborate argument in which he suggests that some Africans, in particular the southern African people he calls "Hottentots," are a nonhuman primate species intermediate between *Homo sapiens,* which he considers to be represented by Europeans, and the orangutan (*Pongo spp.*, although Long appears to be using the word indiscriminately, or ignorantly, to describe any of the Hominidae or great apes). The argument culminates with perhaps his most notorious statement: "Ludicrous as the opinion may seem, I do not think that an oran-outang husband would be any dishonour to an Hottentot female; for what are these Hottentots?—They are, say the most credible writers, a people certainly very stupid, and very brutal. In many respects they are more like beasts than men; . . . that the oran-outang and some races of black men are very nearly allied, is, I think, more than probable."[25]

The passage, like the book in which it appears, has been roundly condemned. "Perhaps more than any other writer in England at this time," argues Roxann Wheeler, "Long looked to African bodies and myths about African life to justify slavery." Peter Fryer describes it as "the classic exposition of English racism," while Srividhya Swaminathan considers it the eighteenth century's "most extreme example of 'racialized discourse.' " Although Long's discussion of race clearly reflects his prejudices, it nonetheless also emerges from a long-standing scientific debate about the nature of species. Long accepts that creation is ordered as a great chain of being. "How vast is the distance," he argues, "between inert matter, and matter endued with thought and reason! The series and progression from a lump of dirt to a perfect human being is amazingly extensive; nor less so, perhaps, the interval between the latter and the most perfect angelic being, and between this being and the Deity himself." In this view, empirical natural history is inseparable from theology; terrestrial organisms are ascribed a place in a chain that continues beyond the observable universe into the realm of speculation and metaphysics. Long's view of the cosmos is conventional in this respect, even though new typologies pioneered by Carl Linnaeus were at this time supplanting the notion of the chain of being, but he is among a growing number of theorists to use the chain of being as a justification for racialized enslavement. In the previous year, in what may have been a source for Long, Samuel Estwick (c. 1736–95), in the second edition of *Considerations on the Negroe Cause,* made a similar case. As part of his critique of Mansfield's ruling, Estwick invokes "that great chain of Heaven" to ask rhetorically if it may "not be more perfective of the system to say, that human nature is a class, comprehending an order of beings, of which man is the genus, divided into distinct and separate species of men?" The argument would become an obsession among later pseudoscientific racists, who continued to order humanity in this way long after the medieval notion of the chain of being had been set aside. In a notorious later example, Charles White (1728–1813), in his 1799 *Account of the Regular Gradation in Man,* argued that "Nature exhibits to our view an immense chain of beings, endued with various degrees of intelligence and active powers, according to their stations in the general system," and while he starts by noting that the chain extends "from man down to the smallest reptile," his claims very rapidly concern themselves only with perceived gradations within the human species. White's *Account* attempted to demonstrate by anatomical comparison that the chain of being operated within "the extremes of the human race" from what he perceived as the highly developed European "to the African, who seems to approach nearer to the brute creation than any other of the human species" and explicitly, "nearer to the ape." Estwick, Long, and White may have been at the extreme, but their views undoubtedly contributed to the development and popularization of racist ideologies that became particularly widespread and pernicious in the nineteenth and twentieth centuries—although there is very little evidence that the polygenist belief that there were separate human species was widely accepted, even in plantocratic circles.[26]

Belief in different human species and in the great chain of being can be used to justify slavery by yoking together taxonomic with scriptural evidence. In Genesis 1:26, God grants humanity "dominion over the fish of the sea, and over the fowl of the air, and over the cattle, and over all the earth." If there is more than one species of humanity, and the European species is highest on the chain of being, then under this reasoning the others are under the dominion of Europeans by the laws of nature and of God. It does not appear that this argument was widely accepted in this period, but Long certainly more than hinted at it. He develops his argument to claim that the further one travels from Africa, the more one observes "gradations of the intellectual faculty, from the first rudiments perceived in the monkey kind, to the more advanced stages of it in apes, in the *oran-outang*, that type of man, and the Guiney Negroe; and ascending from the varieties of this class to the lighter casts, until we mark its utmost limit of perfection in the pure White. Let us not then doubt, but that every member of the creation is wisely fitted and adapted to the certain uses, and confined within the certain bounds, to which it was ordained by the Divine Fabricator." In the concluding lines of his long chapter "Negroes," he reveals the nature of these "bounds" in an ameliorative statement that, while apparently magnanimous, confirms that he sees Africans not only as a subordinate species of humanity but also as one constrained by the laws of nature and of God to a subservient role. "It becomes the gentlemen of Jamaica," he argues, to "conciliate the attachment of their Negroes by protection and encouragement" rather than "by austerity and terror. In the distribution of our gratitude, we are bound to bestow some share on those, whom God has ordained to labour." After 155 pages of prejudice masquerading as science, Long finally comes to the point: God has made two species of humanity, one as slaves and the other as masters. What has been implied throughout is now explicit. Long's apparently highly theorized justification for his nation's enslavement of the people of another, with its polygenism and its chain of being, ultimately rests on a narrow interpretation of natural theology rather than on the broad shoulders of eighteenth-century natural history.[27]

Long's *History* demonstrates a considerable hardening of the position that slavery is natural. His undeniably extreme views on race nevertheless sit within a more conventional plantocratic discourse about the "inevitability" of slavery—that is, that it is a natural phenomenon capable of amelioration but not abolition. This discourse is widely expressed in the mid-eighteenth century, not least by Samuel Martin and James Grainger, who may lament the inability of planter or poet to end the system of enslavement even as they propose methods of paternalistic benevolence intended to make it more profitable. In each case, to lend credibility to the proposition that slavery is natural, or at least naturalized, to the Caribbean environment, the claim must be situated within a wider reading of the landscape that demonstrates the authors' grasp of natural history and their understanding of land management, but that also reconciles modern plantation slavery with traditional practices and traditional modes of representation. Eighteenth-century georgic poets such as John Dyer and James Grainger, argues Christopher Loar, "treat human labor in highly objectionable ways, reducing human workers to factors of production." Grainger, suggests Loar, "flattens distinctions between the human and the material so absolutely that enslaved labor becomes simply another naturalized material in the process of production." What Loar identifies in *The Sugar-Cane* is more broadly the central thrust of mid-eighteenth-century writing that harnessed nature in the service of the plantocracy. Whether presenting their work as island history, plantation manual, or georgic poem, plantation writers represented both introduced sugarcane and imported enslaved laborers as resources now fully naturalized to the Caribbean system, not perhaps beyond management and reform, but certainly beyond abolition.[28]

CHAPTER 4

"The purchase of slaves, teeth and dust"

NATURAL HISTORIES OF THE AFRICAN SLAVE TRADE

Although much farther away, the islands of the Caribbean were better known to seventeenth- and eighteenth-century Europeans than the vast interior of Africa. Nevertheless, travelers did visit Sub-Saharan Africa in increasing numbers from the early sixteenth century onward. Their object was trade, at first mostly for the lucrative commodities of gold dust and "elephants' teeth," or ivory, and their visits were almost entirely confined to the coast and navigable rivers. Increasingly, however, Europeans came to Africa for its people rather than its produce. Scientific inquiry, where it occurred at all, was a sideline rather than a primary purpose. The voyages of John Hawkins (1532–95) and his cousin Francis Drake (1540–96), notorious for introducing the English to Atlantic slave trading, were little more than state-sponsored piracy. The settling of a colony in Barbados in 1627 marked a turning point for English navigation along the West African coast. From this point on, vessels would regularly and increasingly take the fastest route to the island, as Richard Ligon did in 1647, southward to Cape Verde before heading due west to Barbados. The temptation to profit from this part of the journey was irresistible. Ligon's ship took hats to sell at Cape Verde and hoped to purchase horses and enslaved people on the islands. Others did the same, and by the 1660s, the English triangular trade was well established. The addition of further Caribbean colonies augmented maritime traffic. In particular, the capture of Jamaica from the Spanish in 1655–60 led to a dramatic upsurge in English maritime activity in the Caribbean as well as a substantial increase in demand for enslaved laborers. The government of the newly restored King Charles II (1630–85) responded with the formation of the Company of Royal Adventurers Trading into Africa in 1660, reinvented as the Royal African Company in 1672.[1]

The principal purpose of the Royal African Company was to supply enslaved laborers to Britain's American and Caribbean colonies, but it also facilitated an increase in trade with and exploration of Africa that outlasted the company itself and had significant impact on the development of natural history knowledge and collections. In an important study, Kathleen S. Murphy has shown "how naturalists exploited the networks of the British slave trade to acquire seeds and specimens for their museums, gardens, and herbaria," such that science in this period "was inextricably linked with the slave trade." Murphy shows how seventeenth- and eighteenth-century naturalists such as James Petiver (c. 1665–1718) developed an extensive network of overseas collectors, many of whom were officers and surgeons aboard slave ships in the Atlantic trade. Petiver shared his collection and the insights it generated primarily through personal contact and correspondence, and never visited Africa himself. Increasing numbers of voyagers did publish, however, and their narratives fostered public interest in the geography, societies, and wildlife of the continent, generating demand for further reading. As well as accounts by British travelers, translations of voyages by French, Dutch, and other European visitors became ever more popular. The growing volume of publications provided an opening for printers and editors to select and repackage the back catalogue of voyage narratives and issue them in multivolume sets. One of the earliest, and still the best known, was *The Principall Navigations, Voiages, and Discoveries of the English Nation,* issued by Richard Hakluyt (1553–1616) between 1589 and 1600 and containing more than five hundred voyage narratives. Many more would follow. "In the period 1695–1830," Matthew Day calculates, "eighty-five distinct works of this kind were published." Surprisingly, notes Day, although "the link between travel writing and nationalism is well established, the role of travel collections generally in raising nationalist sentiment in the long eighteenth century has not been examined." In fact, Day might simply have asserted that eighteenth-century travel collections have barely been examined for any reason, even though, on average, a new one was published almost every year.[2]

Two collections had a substantial impact on the history of abolitionist writing. The first is the four-volume *Collection of Voyages and Travels,* published in 1704 by brothers Awnsham Churchill (1658–1728) and John Churchill (c. 1663—c. 1714) and posthumously reissued with an additional two volumes in 1732. This is generally known as *Churchill's Collection* (with the apostrophe before the "s," to implying, erroneously, that only one Churchill was involved). The second, and more significant, is *A New General Collection of Voyages and Travels,* compiled by and issued in weekly installments in 1743–45 by Bradock Mead (d. 1757) using the pseudonym John Green, and printed in four volumes by Thomas Astley (d. 1759) in 1745–47. Despite Green's leading role, this is generally known as *Astley's Collection.* Both collections are handsome multivolume sets, Churchill's in folio and Astley's in quarto, with many illustrations. Both boast an impressive list of subscribers, and both outline their methodology in a short preface. Both offer a range of texts written by travelers from across Europe, many of which are offered in original translations commissioned by the editors. They are significant not only for their general contribution to the development and diffusion of eighteenth-century colonial discourse but also because they were Anthony Benezet and Thomas Clarkson's most important sources for African geography, ethnography, and natural history. Although they could have had no idea of it, the Churchills, John Green, and Thomas Astley would therefore assume a disproportionate importance for the development of abolitionist thought and rhetoric. This chapter assesses their representation of Africa before reading the important *Natural History of Senegal,* originally published in French by the naturalist and linguist Michel Adanson, translated into English in 1759, which was likewise widely cited by abolitionist writers. These texts consolidated the view that the African coast presented numerous opportunities for commercial exploitation, but they also presented a multifaceted Africa comprised of varied environments, diverse organisms, and complex societies that were suitable subjects for scientific inquiry—while also sometimes depicting the European demand for slaves as a cruel and unnatural intensification of African cultures and customs.[3]

THE AFRICA OF CHURCHILL'S COLLECTION

Churchill's Collection is the most important voyage anthology of the early eighteenth century, although it is not without flaws and Africa was not originally its editors' main interest. The first edition, published in 1704 and later reproduced as the first four volumes of the second edition, necessarily focuses on sixteenth- and seventeenth-century voyages and contains material mostly already published elsewhere, although much appears in English translation for the first time. Only three of the thirty-four narratives describe voyages to Africa, and these occupy only 154 of the collection's 3,108 pages, or 5 percent of the total. There are six voyages to the Arctic and eight to the Americas, but the Churchills' main interest is Asia, with a total of eleven narratives. The remainder are travels in Europe. This imbalance was noted, and the second edition substantially improved its coverage of Africa. Both editions open in the same way, however, with two substantial essays by the editors, the first historical and the second bibliographical. Each narrative is then separately introduced with a short contextual note, often no more than a paragraph or two, but the text itself, whether entire or an extract, is presented with few obvious editorial interventions.

Two of the Churchills' three African voyages in the first edition are related by Italian Catholic missionaries to Congo, giving the narratives a double layer of exoticism for most British, Protestant readers. The authors present themselves as holy men doing God's business, but their complicity with Catholic colonists and slave traders is never in doubt. Indeed, the church's Congregation for the Propagation of the Faith, often referred to by the short form of its Latin name, Propaganda Fide, was as much concerned with preventing the spread of Protestant traders in Africa, America, and Asia as it was with evangelizing non-Christian peoples. The narrative of Jerom Merolla da Sorrento (1650–1697), whose journey took place in 1682, is a good example. Merolla takes a keen but not entirely reliable interest in wildlife. His account of Congolese fauna reads more like a medieval bestiary than an Enlightenment zoology. He explains how to identify which of an elk's feet is most efficacious against falling sickness (epilepsy); how rhinoceros horns, "being reduced to powder, expel fevers, evacuating by way of sweat the malignity of the distemper"; that "the unicorn, called by the *Congolans, abada,*" has "the same virtue" as the sort of unicorn "commonly mentioned by authors"; and that the wild goat, "if it be eaten when it is lustful, . . . causes such a rot in the feet that the toe-nails drop off." His medically inflected bestiary leads him to consider another African phenomenon. " 'Tis now high time to leave the wild beasts to range in the woods," he asserts, "and to come to speak of a certain brutish custom these people have amongst them in making of slaves, which I take not to be lawful for any person of a good conscience to buy here." It is not, however, the making of slaves itself that Merolla considers unlawful but only the "brutish custom" by which enslaved women are ordered by their enslavers to seduce young men who are then themselves enslaved in fake revenge by the enslavers, pretending to be the women's outraged husbands. Merolla also objects to "others who, not by means of women but of themselves, going up into the country thro' pretence of jurisdiction, seize men upon any trifling offence, and sell them for slaves." Again, it is the method of enslavement rather than slavery itself that Merolla condemns, and even then the "brutish custom" is depicted as an organic part of the African environment, naturally and immediately following on from discussion of "brutes" such as the rhinoceros, elk, and unicorn.[4]

This contiguity of ideas might be dismissed as a coincidence, except that Merolla precisely repeats the maneuver fifty pages further on. At the conclusion of another long description of wild and often fantastical animals, he abruptly switches "to give an account of more memorable matters." Again, these matters concern slave trading. Merolla is sent a letter from "Cardinal Cibo" asserting "that the pernicious and abominable abuse of selling slaves, was yet continued among us, and requiring us to our power to remedy the said abuse." As Richard Gray has shown, the letter was part of a general condemnation of the slave trade issued by the Vatican at this time, albeit with little lasting effect. Merolla's response is lukewarm. His concession is to convince the king of Congo that "the hereticks at least should be excluded from dealing in this merchandize; and that especially the *English,* who made it their chief business to buy slaves here, and to carry them to *Barbadoes,* an island of theirs in the *West-Indies,* where they were to be brought up in the protestant religion." When an English vessel appears, Merolla tries to enforce this, and the English captain counters by reminding him "*Is not our duke of* York *a* Roman *catholic, and chief of our company?*" The dispute grows vexatious and the king intervenes to override Merolla and continue slave trading with the English. Merolla's apparent belief that slavery was acceptable under Catholic but not Protestant enslavers appears confused at best. Indeed, Gray describes it as "a narrow and mistaken interpretation of Propaganda's new intentions." It would probably have seemed hypocritical to English readers in the early eighteenth century, albeit only because English readers would have thought it unreasonable to be excluded from the trade. The captain's reference to the Duke of York would have occasioned some irony, given that James, Duke of York (1633–1701) went on to become king of England before being ousted in the revolution of 1688. Nevertheless, while Merolla depicts the English captain as a blowhard who tries "to overcome reason with noise," readers in England may have identified him as a plain dealer, in the mold of Captain Manly from William Wycherley's (1641–1714) play *The Plain Dealer* (1676).

For English readers, the episode reveals Merolla as duplicitous and Jesuitical, and the Vatican's foreign policy as self-serving and treacherous.[5]

The book's primary interest in 1704 was not, however, its account of either slavery or Catholic missionary politics but rather its depiction of the African continent. Merolla shortly after continues to observe and describe African wildlife in vivid detail, albeit without any adequate vocabulary or system. His long descriptions of the mermaids and seahorses in the Congo River probably describe the African manatee (*Trichechus senegalensis*) and the hippopotamus (*Hippopotamus amphibius*) but demonstrate that he is attempting to describe Africa through the lens of preexisting knowledge acquired through reading rather than personal experience. The same is true when he talks of African magicians, wizards, electors, and lords of the manor. John Green, the editor of *Astley's Collection,* saw the inclusion of such material as a major defect of *Churchill's Collection.* The narratives it contained, he believed, "*seem to have been gathered without Judgement or Care,*" while some "*are swelled with scarce any Thing but the Transactions, and even the Disputes, of Missionaries.*" The Churchills may, however, have judged that the transactions of Catholic missionaries, their appearance of hypocrisy in the face of honest English plain dealing, and their antiquated method of describing African flora and fauna, sufficiently discredited both Catholic knowledge systems and Catholic colonialism to resonate with a Protestant English audience, and this might explain why these accounts were included despite their obvious deficiencies.[6]

A second edition was printed in 1732 "by Assignment from Mss^rs^ CHURCHILL," both of whom had recently died. Almost half of the additional new pages are given over to just two accounts of voyages to West Africa. The shorter of the two is an account by one Thomas Phillips of "a trading voyage to *Guiney,* for elephants teeth, gold, and *Negro* slaves" made in 1693. This contains considerable detail of day-to-day life aboard a slave ship but has relatively little to say about the African continent itself, which Phillips mainly saw only from his ship or from European outposts. The longer of the two fills almost all of the new fifth volume. *A Description of the Coasts of North and South Guinea* was crafted by the publishers from the manuscripts of Jean Barbot (1655–1712), a French Huguenot slave trader who fled from France to England after the Revocation of the Edict of Nantes in 1685. The *Description* would become an important source text for African history even though, as Robin Law has pointed out, "the problems involved in its use as a historical source have been recognized for some time." These are, principally, that Barbot made extensive alterations to his manuscript between starting it in French in the 1680s and completing it in English before his death in 1713, that the publishers extensively revised his manuscript before publication in 1732, and that both he and the publishers incorporated much material from others, particularly from *A New and Accurate Description of the Coast of Guinea,* written by the Dutch merchant Willem Bosman (b. 1672) and published in English in 1705. This important and authoritative text, which explicitly tackles African natural history, is referred to throughout the eighteenth century and beyond in discussion of Africa, but often via its incorporation into other works, including both Churchill's and Astley's collections. Indeed, in his edition of the 1688 manuscript of Barbot's account of West Africa, Paul Hair notes that "a major problem in editing Barbot is the recognition and attribution of derived material," not least because of Barbot's "reluctance . . . to state his sources." These issues are certainly challenging for historians but they do not necessarily diminish the rhetorical force of the text, nor do they impact on the ways in which Barbot was appropriated by later abolitionist writers. Although problematic as a social, economic, or natural history of West Africa, the text offers significant information about the late seventeenth-century slave trade and important insights into eighteenth-century knowledge about and attitudes toward Africa. We may, in fact, more profitably regard *A Description of the Coasts of North and South-Guinea,* as it appeared in *Churchill's Collection* in 1732, as a unified composite text with no single author since this is the way it would have been experienced by its eighteenth-century readers.[7]

At around half a million words, the *Description* is too long to consider in depth, but unlike the missionary accounts included in the first edition of *Churchill's Collection,* the narrative is simultaneously a natural history, an ethnography, and a guide for slave traders. The title page advertises itself as containing "A Geographical, Political, and Natural History" of the region while also explicitly offering "Measures for improving the several Branches of the *Guinea* and *Angola* Trade." In his eclectic "Introductory Discourse," which is apparently his own work, Barbot advises the visitor to West Africa to go both as a humanist and as a naturalist, fluent in numerous languages, both African and European, but also with "some skill in drawing, and colouring, that he may be able to take draughts of prospects, landskips, structures, birds, beasts, fishes, flowers, fruits, trees, and even of the features and habits of people." In his insistence on the importance of careful observation and recording, Barbot identifies himself as a dedicated naturalist and not merely a slave trader who chanced to take a few notes along the way. This is, for him, an important distinction. He early on laments how few traders return from Africa with any useful knowledge of the continent, despite their being "scarce any other voyage that will afford a man more leisure to observe and write, whether he goes only on a trading voyage, or resides there; because there is not always a brisk trade, so that every man may have spare hours to make his remarks, and write them down as they occur." Barbot is aware that he is an exception in making good use of his time with natural history. "Two or three hours every day," he reasons, "may be better employ'd that way, than in drinking, gambling, or other idle diversions too frequently used." The "Introductory Discourse" thus asserts its authority in the moral as well as the physical sphere.[8]

We may reasonably safely speak of Barbot's authorial presence in the introduction, but much of the 1732 *Description* conflates Barbot's voice with those of Bosman, the publishers, and several others. I call this composite narrator "the Describer" to distinguish him from the historical Barbot. The 1688 Barbot emerges as a curious traveler, a keen observer, and a vivid writer with an interest in wildlife and environment as well as trade and society. He demonstrates, however, no specialist knowledge of typology, botany, or zoology. The 1732 Describer, by contrast, is presented as a gentleman naturalist with an awareness of the formal conventions of the natural history genre, often signaled structurally by subheadings and side notes. Sometimes, like many of the gentlemen and clerical naturalists who depicted the Caribbean, the Describer presents information that is impressionistic, incomplete, or imprecise. The forests around the Senegal River, for example, described in chapter 2, "harbour prodigious numbers of elephants" alongside "lions, leopards, tygers, rhinocerots, camels, wild asses, wolfs, wild goats, stags, ounces, panthers, antelopes, fallow deer, wild rats, wild mules, bears, rabbits, and hares." These coexist with "very many apes, monkeys, and baboons" as well as "a prodigious number of extraordinary large lizards, which are good to eat." In these pages, the Describer repeatedly emphasizes the plentitude and culinary virtue of West African fauna (and elsewhere flora) with little attention to detail. The exceptions are when the animal is mysterious or unusual. The Describer comments at greater length on "a sort of wing'd, or flying serpent, which uses to feed on cow's milk, sucking it at the dug, without hurting the beast." Such descriptions verge on the mythological. Others are more recognizably confused accounts of novel animals, such as a creature with "the body of a dog, and the hoofs of a deer, but larger, the snout much like that of a mole, and feeds on ants, or pismires; and, if we may believe the *Blacks*, digs as fast with that snout under ground, as a man can conveniently walk." This chimera, illustrated on the following page, is almost certainly the aardvark (*Orycteropus afer*). The description, found also in Barbot's 1688 manuscript, tells us that Barbot sought out local knowledge to interpret novel environments but also that his interactions with Africans were marked by degrees of skepticism and distrust.[9]

Elsewhere, however, the *Description* adheres more closely to the formal and descriptive conventions of natural history. The Describer's account of the hippopotamus in the Gambia River, for example, is detailed, extended, and careful to separate observed features from hypothetical attributes. The animal's skin is "said to be good against the looseness and bloody-flux," while its teeth "are said to have a physical virtue to stop bleeding, and cure the hemorrhoids, as has been found by experience." This careful distinction between hearsay and experience appears increasingly throughout the *Description*—although, as Hair points out, this passage, which also appears in the 1688 manuscript, "must be either from oral information or from an untraced printed source" since Barbot himself "never visited River Gambia." As the book progresses, the descriptions become more detailed and more formalized. The chapter on the "Trees, animals, birds, and insects" of Quoja (the Koya Kingdom in present-day Sierra Leone) lists twenty-two species of tree and a dozen species of large animal, each with a separate heading and a short description. A separate section divides insects into two sorts, vipers and serpents, followed by separate headings for twenty-three species of bird, including the bat. Flying insects are divided into honey bees, gnats, and "a multitude of flies." The typology may seem eccentric in the decade in which Linnaeus published the first edition of the *Systema Naturae,* but it is a typology nonetheless, not merely a list of nature sightings. Further on in the book, the wildlife of the Gold Coast (present-day Ghana) is described with even more precision and detail, but here the text is lifted almost verbatim from Bosman: an extended piece of plagiarism that must have been apparent to any contemporary reader of both books. Nevertheless, the *Description* becomes increasingly comprehensive and authoritative as a natural history of the coast of West Africa. The multiple textual sources focused through the lens of Barbot produce an image of the region as exotic, biologically diverse, and highly productive. This is supported by illustrations that crowd together people, wildlife, and agricultural produce; a good example is the plate depicting the Sestro River (today the Cestos River in Liberia), where the Describer went ashore to meet "king Barsaw, an elderly man, with silver hair" (figure 4). Its six panels include a map, three species of endemic fish, two species of bird, about half a dozen plants, an ant, a worm, a monkey, a sheep, a village, and a family group. The intention may have been purely scientific, but the effect is reminiscent of a commercial catalogue of goods available for purchase.[10]

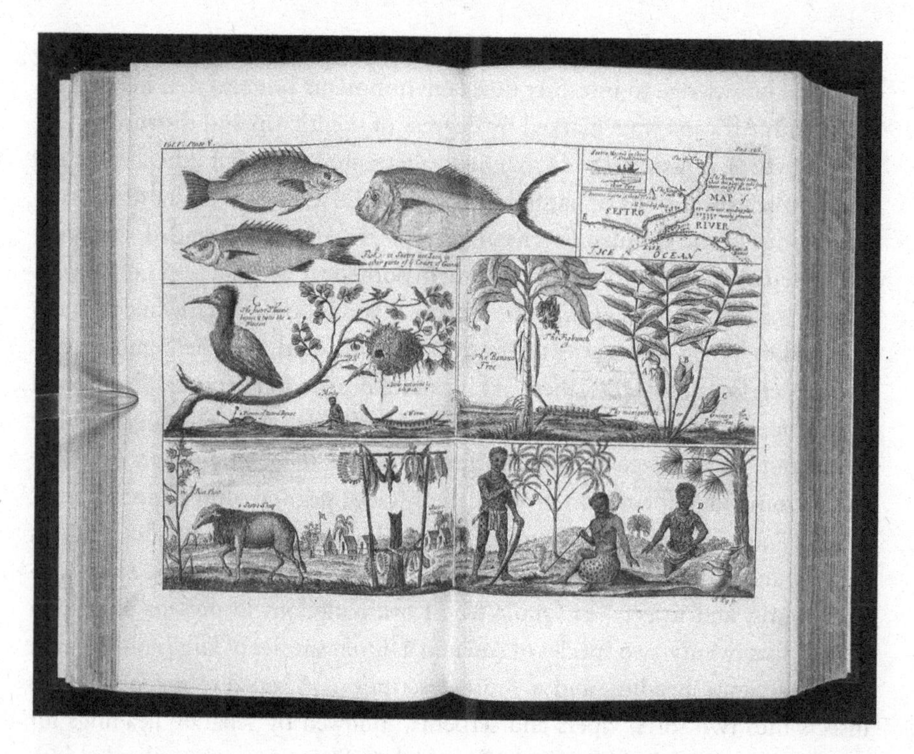

Figure 4. Six-panel illustration of the flora, fauna, and people of the Sestro River area. From *A Collection of Voyages and Travels,* 2nd ed., 1732. (© The British Library Board.)

Africa's plentitude is, indeed, the overriding impression that emerges from the *Description.* Whenever the continent's flora or fauna are mentioned, the Describer goes overboard with superlative and hyperbole. In Sierra Leone, for example, the country is "overrun with lofty trees," the forests "harbour infinite numbers of parrots," and "the seas and rivers furnish the natives and travellers with abundance of fish." This fecundity is not always desirable. The country "swarms with elephants, lions, tygers, wild boars, fallow and red deer, roes, apes of several sorts, and serpents: some of these last, so monstrous big, if we may credit the natives, that they swallow a man whole." As well as fish, the river banks "are all hemm'd in with mangrove-trees, on which stick abundance of oisters." These "produce very fine pearls; but it is very dangerous taking of them, because of the infinite number of sharks." Nevertheless, the land itself, which has "excellent fresh water" and "abounds in rice, maiz, ignames [yams], bananas, citrons, oranges, pompions [pumpkins], water-melons, and the fruit *Cola,*" is benign enough to come under the Describer's imperial gaze. The coastal islands, he notes, produce "abundance of oranges, lemons, palm-wine, and sugar-canes growing wild, which is a demonstration that the soil is proper for sugar plantations; besides, that there are many brooks and springs which would turn the mills at a cheap rate." The observations of a naturalist are quickly turned to commercial account.[11]

The superlative tone is repeated in numerous passages. These describe both cultivated and uncultivated land and both wild and domesticated animals. In places, description of the land leads into a description of African society that is equally emphatic about its wealth and diversity. In his account of the nations along the Gambia River, for example, the Describer notes the "excellent pasture grounds, which serve to feed immense herds of cattle." This is no monoculture: "The country is also well stored with goats, sheep, elephants, lions, tygers, wild boars, and many other sorts of tame and wild beast; especially about *Mansagar,* where they have great droves of horses, camels, and asses." The birdlife is equally remarkable. "As for poultry," he notes, "the plenty is incredible; and so of parrots and parrokeets, with many other sorts of birds, several of them very remarkable for the wonderful variety and beauty of their feathers." The only disadvantage is the "unwholesome" air that "is occasion'd by the malignant vapours rising from the marshy grounds and thick woods and forests." This is compounded by "intolerable heats," "excessive rains," and "horrid thunder, lightning, and tornado's." Like the land, the climate itself is prodigious, but this balances the fecundity of the land and checks European ambitions in the region. "Were it not for this destructive disposition of the air," he reasons, "it might be pleasant living in that country, being so fertile and good." Europeans may still exert their influence, however. The passage immediately shifts from discussion of the land to description of the people, and the Describer is in no doubt that European influence is a positive force in the region. "THE *Blacks* of *Gamboa,*" he argues, "were formerly very savage, cruel and treacherous; but through long commerce with the *Europeans* they are now become pretty tractable; especially those about the sea-coasts, who are most civiliz'd." His description of the transition from savagery to civility forms the textual bridge between his descriptions of African environment and African society and implies that both commerce and civilization are stimulated by contact with European traders. "The fair at *Mansagar,*" he notes, "is held under a hill, near the town, where some Portuguese Mulattoes have their dwelling." As with the land, villages along the Gambia River are remarkable for the quantity, richness, and diversity of their produce: "Thither is brought to the market abundance of salt, wax, elephant's teeth, mats, cotton, gold-dust . . . all sorts of cattle, goats, poultry, horses." At nearby Great Cassan, "an almost incredible number of people" visit the fair, but because the river is not navigable, "all the goods they carry thither, or bring back, must go and come by land on the backs of slaves."[12]

This is a note in passing. Enslaved people figure surprisingly sparsely in a narrative based on an account written by a slave trader. Throughout, the book essentially accepts that slavery is a natural phenomenon not requiring detailed description. The passages that do discuss the slave trade promote the view that enslaved people "are, for the most part, people taken in war," although a few sell themselves into slavery to relieve themselves from debt, some are "sold into bondage by their own relations," and others "are sometimes stolen away, out of their own countries by robbers." The slave trade in Africa is "the business of kings, rich men, and prime merchants" and "these slaves are severely and barbarously treated by their masters, who subsist them poorly, and beat them inhumanely." Despite the ostensibly civilizing influence of European trade more generally, the trade in slaves is thus portrayed as a natural consequence of preexisting African power struggles rather than as a result of European demand. A long description of "the management of slaves" later in the book attempts to represent the European trade in enslaved Africans as a separate and more humane business than the internal African slave trade. The Describer assures the reader that slave traders are "very nice in keeping the places where the slaves lie clean and neat," and that "thrice a week we perfume betwixt decks with a quantity of good vinegar." He positions himself as an exemplary trader. "We allow'd them much more liberty," he tells us, "and us'd them with more tenderness than most other Europeans . . . few or none being fetter'd or kept in shackles." On the other hand, he notes, "We keep all our small arms in a readiness, with sentinels constantly at the door." The slave ship is, after all, still a place of incarceration. The passage concludes with an exhortation that would not have been out of place in a Quaker antislavery text of the period. Officers of slave ships, he argues, "should consider, those unfortunate creatures are men as well as themselves, tho' of a different colour, and pagans; and that they ought to do to others as they would be done by in like circumstances; as it may be their turn, if they should have the misfortune to fall into the hands of *Algerines* or *Sallee* men, as it has happen'd to many." The Golden Rule and the fear of capture by Muslim sailors were both notable features of the nascent antislavery writings of Quakers before 1732, and it is possible, if unlikely, that some of these publications might have fallen into the hands of either Barbot or the Churchills. Either way, the *Description* advertises the slave trade as a humane branch of a naturally occurring business, arising from an African society that itself emerged from a continent replete with both natural and human resources.[13]

This is also the overwhelming impression of Africa that one receives from reading the 1732 edition of *Churchill's Collection* in its entirety, not least because the *Description* is the longest single text in the collection. The African narratives, from the accounts by Italian missionaries to those of the slave traders Phillips and Barbot, offer different angles on essentially the same message, even if the message is sometimes somewhat inconsistent. All of them consider it important to describe the flora and fauna of Africa, whether simply as curious visitors or as would-be natural historians and, albeit to different extents, all of them show more interest in African wildlife than they do in the slave-trading business that brought them to the continent. They all, perhaps inevitably, portray African wildlife as exotic and diverse but this is within a broader framework that depicts the continent itself as highly productive—indeed, as prodigious. Within this frame, African society is represented as complex, thriving, and diverse, even if Africans as individuals are frequently portrayed as unchristian, uncivilized, and untrustworthy. The authors unite in portraying themselves as honest and humane slave traders even when expressing regret or even disquiet about the broader humanity of slavery. All, however, see slavery as an inevitable consequence of wars and other naturally occurring internal conflicts. Readers of *Churchill's Collection,* then, would have encountered an Africa celebrated largely for its prodigious natural resources, which included spices, gold, and ivory as well as enslaved people.

THE AFRICA OF ASTLEY'S COLLECTION

A decade after publication of the second edition of *Churchill's Collection,* the Irish cartographer Bradock Mead (d. 1757), usually known by his pseudonym John Green, compiled *A New General Collection of Voyages and Travels.* This massive compendium of around 2.5 million words was issued in weekly installments in 1743–45 and printed in four volumes by Thomas Astley in 1745–47, giving rise to the name *Astley's Collection,* by which it is generally known. *Churchill's Collection* had offered extracts and full texts, supported with some brief contextualizing introductions, but had not otherwise meddled substantially with the texts themselves. Green takes a different approach, prompted at least in part by his apparent outrage at the Churchills' method. *Churchill's Collection,* argues Green, "*is no more than an Assemblage of the*

Travels of about fifty particular Authors to a few Parts of the World" while "*the Authors made use of are, for the most Part, of very little Esteem. They seem to have been gathered without Judgment or Care; and chosen (if there was any Choice made at all) rather for their imperfections than Merit.*" Green is critical that "*the foreign Authors are very badly translated,*" but he is prepared to concede that "*the Part best executed in the Whole, though very dry, is the Introduction.*" There is a certain amount of performative commercial rivalry in this, and Green in fact makes use of several of the Churchills' imperfect authors in his own collection, but he also uses his critique of their method to justify his own unique selling proposition. Hoping to cram more information into a shorter space, he tells the reader that "*we have deviated from the common Method of collecting, and instead of giving each Author entire in the Order he was published, we separate his Journal and Adventures from his Remarks on Countries: The first we give by itself; the latter we incorporate with the Remarks of other Travellers to the same Parts.*" This plan to separate the maritime derring-do from the geography, ethnography, and natural history allows Green to play (and to sell) to two audiences, but it also has the effect of transforming a heterogenous set of voyage narratives into a coherent piece of environmental science. "*And thus,*" proclaims Green, "*our Collection becomes a System of* Modern Geography and History, *as well as a Body of Voyages and Travels.*"[14]

This transformation reflects the interests of a period that we might call the heroic age of reference works. *Astley's Collection* appeared at a time when Samuel Johnson (1709–84) and Denis Diderot (1713–84) were at work on the *Dictionary* and the *Encyclopédie* respectively. Green himself had served as an amanuensis to Ephraim Chambers (1680–1740) during production of the *Cyclopædia,* published in 1728, contributing the entries on "map" and "projection." Bringing together disparate knowledge into a complete system was one of the characteristic intellectual activities of the eighteenth century, but Green's approach was especially important for the future of abolitionist writing, given that *Astley's Collection* was a major source of information for Anthony Benezet and, through him, the many abolitionist writers who drew on his research. Indeed, notes David Crosby, *Astley's Collection* "is by far Benezet's most important source. He mined the first three volumes for 90 percent of the quotations in his first two pamphlets." Like *Churchill's, Astley's* was popular and widely distributed—Benezet had access to a copy in colonial Pennsylvania—but it was far more comprehensive even if the extracts it offers are often heavily abridged. Where *Churchill's* had offered only six narratives that described Africa, *Astley's* contains extracts from more than a hundred.

The whole of volume 2 and large sections of volumes 1 and 3 are dedicated to African voyages. In total, 1,425 of the collection's 2,768 pages—just over half—are concerned with Africa, although the final proportion might have diminished had Green not fallen out with Astley before he was able to edit the volumes on America, which never appeared. Nevertheless, the authors represented are numerous. Among others, the book includes extracts from African narratives by John Atkins (1685–1757), Jean Barbot, André Brüe (1654–1738), Richard Jobson (fl. 1620–23), Francis Moore (c. 1708—c.1756), Thomas Phillips (c. 1664–1713), and William Snelgrave (1681–1743).[15]

It is useful to follow Green's own lead and assess the representation of Africa offered in the "journal and adventures" separately from the "remarks on countries." Unlike the "remarks," the "adventures" are recognizably based on extracts from individual authors, albeit heavily modified. Green introduces each author with a contextualizing introduction, but the texts themselves are abridged, altered from a first-person to a third-person narration, and frequently fragmented. With hundreds of extracts to choose from, selecting a representative text is almost an arbitrary decision. Nevertheless, Green's deployment of John Atkins's *Voyage to Guinea,* originally published in 1735 and repackaged in several extracts scattered across volumes 1 and 2 of *Astley's Collection,* provides a useful case study. Atkins was a Royal Navy surgeon who joined an anti-piracy mission to West Africa in the early 1720s. His account of the voyage is notable for its application of Newtonian physics to meteorology; its attempt to describe what today we call earth systems, in particular the water cycle; and its skepticism concerning travelers' tales about cannibalism. It expressed considerable disquiet about slave trading while also positing a polygenist theory of race. Atkins found conventional climatic explanations for race unsatisfactory and argued that "tho' it be a little Heterodox, I am persuaded the black and white Race have, *ab origine,* sprung from different-coloured first Parents." Atkins's was a serious early attempt to explain differences in skin color, not an attempt to seek a racial justification for slavery, as was the case with later polygenists such as Edward Long. Indeed, alone among all the voyagers in *Astley's Collection,* Atkins is included in Thomas Clarkson's 1808 map of "Abolitionist Forerunners" (see figure 1) because his *Voyage to Guinea* "describes openly the manner of making the natives slaves, such as by kidnapping, by unjust accusations and trials, and by other nefarious means." Many passages in Atkins's original text express disquiet or disgust at the activities of European slave traders, several of which are included in *Astley's Collection.* Much of the text is, however, truncated or dispersed. Green's treatment of Atkins, and by extension his treatment of many of his primary sources, can be illustrated by placing Atkins's original opening paragraph next to Green's version:

John Atkins, *Voyage to Guinea,* 1735

We took in eight Months Provisions each, at *Portsmouth;* Stores, Careening-Geer, and Necessaries requisite to continue us a double Voyage down the Coast of *Guinea,* for meeting, if possible, with the Pyrates; who did then very much infest those Parts, and destroy our Trade and Factories. Accordingly the Company's Governors for *Gambia* and other Places, embark'd under our Convoy, and were to have what Support we could give them, in restoring the Credit of the *Royal African Company;* which begun now to take new life under the influence of the Duke of *Chandois.* For this Purpose we set sail from *Spithead February* 5th $17^{20}/_{21}$

Atkins in *Astley's Collection,* 1745

They sailed from *Spithead, February* the fifth, 1720–21, taking in Necessaries for a double Voyage down the Coast of *Guinea,* with an Intention to destroy the Pirates, who greatly infested those Parts, and destroyed their Trade and Factories. The *African*-Company Governors for *Gambra* and other Places went under their Convoy.

Not only is Green's version half the length of the original, and changed to a third-person narration, but it has been reordered, key information has been omitted, and place-names have been changed. Confusingly, even though this extract appears to begin at the beginning, and is prefaced with Green's contextual introduction to Atkins, it is in fact the third time a substantial extract from Atkins's *Voyage* appears in *Astley's Collection;* the previous two are left unintroduced. Nevertheless, Green's version of the narrative is pithier and more focused, which apparently found favor with some readers. "*Some of our Subscribers have complained,*" notes Green in the introduction to his second volume, "*that they thought we curtail the Adventures of Travels too much in our Abstracts; and yet others have imagined that we did not retrench them enough: The Truth is, we have endeavoured to avoid both Extremes.*" Green accordingly invites the reader to compare the originals with the abridgements and to agree with him "*that we have given the Substance of every Thing that seemed any way material. Had we inserted every Thing, which those Authors thought fit to insert in their journals, we should perhaps have nauseated even those who read Voyages and Travels chiefly for Sake of the Adventures.*"[16]

Throughout *Astley's Collection,* the "Adventures" contain numerous accounts of prosperous gales, mountainous seas, and treacherous reefs. They depict meetings with indigenous peoples as well as other European travelers. They recount encounters with hostile warships, rapacious pirates, and commercial rivals. Indeed, for eighteenth-century readers, the word *adventure* did not mean only an exciting journey but also a business venture, and both traders and investors might be referred to as "adventurers." The Royal African Company itself had started out as the Royal Adventurers into Africa. Thus, the adventures reproduced in the African narratives in *Astley's Collection* are very often accounts of slave trading. Again, Atkins is a useful test case. He had discussed the slave trade at length in *Voyage to Guinea,* and much of this is carried over into *Astley's Collection.* The first discussion comes in Atkins's account of Sierra Leone, reproduced with relatively little abridgement. Atkins draws a contrast between the cruelty and venality of the European "private traders" and the nobility of one of the Africans whom they have enslaved. These traders, described by Green as "Buckaneers" in a corroborating footnote, are "loose, privateering Blades, who, if they cannot trade fairly with the Natives, will rob; though not so much to amass Riches, as to put themselves in a Capacity of living well and treating their Friends." The worst of this crew is "*John Leadstine,* commonly called *Old Cracker.*" All of them "keep *Gromettas* (or Negro Servants) whom they hire from *Sherbro* River." The women, says Atkins, "are obedient to any Prostitutions their Masters command." The pirates' hired hands take British metal goods and Caribbean spirits upriver in exchange for "Slaves and Teeth." Atkins describes the sale of the ivory and the captives arriving in chains. "Most of them," he notes, with some understatement, "were very dejected." Most, but not all:

> Once looking over some of *Old Cracker*'s he took Notice of one who was of a tall, strong Make, and bold, stern Aspect. This Fellow seemed to disdain the other Slaves for the Readiness to be examined, and scorned to look at the Buyers, refusing to rise or stretch-out his Limbs as the Master commanded. This got him an unmerciful Whipping, with a cutting *Manatea*-Strap, from *Cracker*'s own Hand; who had certainly killed him, but for the Loss he must have sustained by it. The Negro bore it all with Magnanimity, shrinking very little; but shed a Tear or two, which he endeavoured to hide, as though ashamed of.

> The Company, upon this, being curious to know how *Cracker* came-by him: He told them, that this Person, called Captain *Tomba,* was a Leader of some Country Villages which opposed them and their Trade, at the River *Nunes,* killing their Friends their and firing their Cottages: That the Sufferers, by the Help of his (*Cracker's*) Men, having surprized him in the Night, about a Month before, brought him thither; but that he had killed two of them, in his Defence, before he was taken and bound.

Hugh Thomas notes that this account represents "a rare example of an African ruler seeking to prevent or at least to resist the slave trade." Although rare, it nevertheless conforms to the literary conventions common to European depictions of African resistance to enslavement, whether in Africa or elsewhere, and many would have recognized Tomba as a compatriot of Oroonoko. He is portrayed as a warrior prince, valiant in battle, dignified in defeat, demonstrating great stoicism in the face of suffering but nonetheless betraying a deep sensibility, marked by the tear that that he attempts to hide. The passage clearly invites the reader to condemn Cracker and sympathize with Tomba, but Atkins leaves it without further direct comment. Instead, he immediately turns to the natural history of the Sierra Leone River and, in particular, the African manatee whose hide had supplied the strap Cracker used to whip Tomba. Atkin's description of the animal is precise and detailed, noting, as was common in the period, its culinary and other uses as well as its morphology. He pays particular attention to the cuticle. This, he says, is "granulated, and of a Colour and Touch like Velvet; the true Skin is an Inch thick, and, by the *West Indians,* used in Thongs for punishing their Slaves." Readers may have recalled Hans Sloane's discussion of the manatee-hide whip in *A Voyage to Jamaica.* Unlike Sloane, Atkins does not comment on whether these punishments are "merited" or not but, by placing the natural history of the manatee in such close proximity to the story of Tomba, Atkins allows the reader to infer that it is not.[17]

We might be cautious of interpreting Tomba's story as an unambiguous statement of antislavery since Cracker was a privateer, living beyond the law. Tomba's resistance continues, however. He is sold to a legally sanctioned slave trader, Captain Richard Harding, commander of the *Robert* of Bristol. On board the *Robert,* Tomba leads an insurrection in which he and four others, one a women, are overpowered. The outcome is horrific. Harding sentenced three of the men "to cruel Deaths, making them first eat the Heart and Liver of one of them he killed. The Woman he hoisted by the Thumbs, whipped, and slashed her with Knives before the other Slaves, till she died." Green adds the side note "Devilish Cruelty" but omits Atkins's several direct statements of opposition to the slave trade, some of which were a direct response to William Snelgrave's arguments in favor of the trade that had been published the previous year. Atkins contended that the slave trade contravened natural justice. "An extensive trade," he had argued in his original text, "in a moral sense, is an extensive Evil, obvious to those who can see how Fraud, Thieving, and Executions have kept pace with it. The great excess in Branches feeding Pride and Luxury, are an Oppression on the Publick; and the Peculiarity of it in this, and the Settlement of Colonies are Infringements on the Peace and Happiness of Mankind." Later, he concluded, "To remove *Negroes* then from their Homes and Friends, where they are at ease, to a strange Country, People, and Language, must be highly offending against the Laws of natural Justice and Humanity; and especially when this change is to hard Labour, corporal Punishment, and for *Masters* they wish at the D——l." Should any of his readers think this an abstract or distant consideration, Atkins insists that "we are Accessaries by Trade, to all that Cruelty." Sentiments such as this justify Clarkson's inclusion of Atkins among the abolitionist forerunners but Green, while happy to reproduce the story, or "adventure," of Captain Tomba in *Astley's Collection,* clearly drew the line at reproducing Atkins's misgivings about the slave trade. Harding's personal act of "devilish cruelty" might be revealed in his collection, but it would be up to the reader to extrapolate the cruelty of a single captain to the entire slave trade. Such editorial reservations did not, however, apply to the opposite point of view. Green is content to reproduce William Snelgrave's defense of the slave trade at length.[18]

Despite this apparent double standard, and perhaps sensing the importance of the juxtaposition, Green reproduced the natural history of the manatee just as it had appeared in Atkins's original account: immediately after the account of Cracker assaulting Tomba with his manatee-hide whip. In general, however, he separated out into discrete chapters the "remarks on countries" in which natural history is collated. These "remarks" are complex and fragmented. In some parts of the collection they combine topography, ethnography, and natural history, while in others these topics are carefully differentiated. Although at the level of chapter headings Green treats society and nature as separate spheres, in reality he finds it difficult to detach the activities of people from the environments in which those activities take place. Discussions of both European and African slave trading are relatively frequent in the mostly ethnographic chapters but less so in the chapters largely dedicated to natural history. In the former, for example, there is an entire chapter, "*Of the* English *Trade on the* Gambra," largely drawn from Francis Moore's *Travels into the Inland Parts of Africa,* printed by Green's former employer Edward Cave (1691–1754) in 1738. This combines observations on the operational details of the Royal African Company with Moore's critique of the effect of slave trading on African society. Moore notes that most enslaved Africans are "Taken in War, . . . or Men condemned for Crimes; or Persons stolen, which is very frequent." Many of the criminals were condemned for minor crimes. "Since the Slave-Trade has been introduced," says Moore, "all Punishments are commuted into this; and they strain hard for Crimes, in order to have the Benefit of selling the Criminal: So that not only Murder, Theft, and Adultery, but every trifling Crime is, at present, punished with Slavery."[19]

This account is critical of the Royal African Company's influence on African society, but it does not exonerate Africans from complicity in the trade. Indeed, throughout these sections, Africans are repeatedly represented as, in Barbot's words, "lewd and lazy to Excess, which makes them miserably poor; impudent, knavish, revengeful, proud, and fond of Praise; extravagant in their Expressions, Liars, abusive, gluttonous, extremely luxurious, and so intemperate, that they drink Brandy like Water; fraudulent in dealing: Rather than work, they will rob and murder on the high Way, or carry-off those of a neighbouring Village, and sell them for Slaves." African societies, cultures, and customs are frequently shown as being constructed around these personal qualities. Whether pseudoscientific or not, this is racism, and it reflects the increasingly prevalent use of racial stereotypes to justify the slave trade. Nevertheless, throughout the ethnographic passages in Green's collated "remarks on countries," slavery, while represented as an organic part of African societies, is also repeatedly depicted as a part that is being distended beyond its natural limits. To Moore's comments, Green adds those of the French ethnographer Jean Baptiste Labat (1663–1738), whose 1728 *Nouvelle relation de l'Afrique occidentale* was itself a composite collection. Moore acknowledges African slavery, noting that some "have a good many House-Slaves, in which they place a great Pride." By contrast, in Green's translation, Labat recounts the effect of European demand on the African slave market. "The King furnishes the *Europeans* with Slaves very easily," says Labat. "He sends a Troop of Guards to some Village, which they surround; then seizing as many as they have Orders for, they bind them up and send them away to the Ships, where the Ship-Mark being put upon them, they are heard of no more."[20]

Astley's Collection offers somewhere between a quarter and half a million words of African ethnography. Its natural history is only slightly less extensive. In volume 2, natural history appears in seven contiguous chapters describing voyages from "Cape Blanco" (modern-day Ras Nouadhibou in Mauritania and Western Sahara) to Sierra Leone, and later in a single chapter on the Gold Coast (present-day Ghana). Volume 3 continues the pattern, with a single chapter covering the natural history of the coast from "Rio da Volta to Benin" (approximately modern-day Nigeria), another describing "Kongo, Angola, and Benguala" (essentially, modern Angola), and another the whole coast of East Africa from "Cape of Good Hope to Cape Guarda Fuy" (Cape Guardafui in modern-day Somalia). Other parts of the African coast are omitted or included in more general accounts. Again, the scale of the publication makes it difficult to assess except through a case study. The seven chapters in volume 2 covering Mauritania to Sierra Leone are both the lengthiest and the most representative, describing the part of the Sub-Saharan West African coast closest to Europe and thus most visited by European voyagers. In these, Green divides his organisms "into five Classes, *viz.* the Vegetables, Quadrupeds, Birds and Fowl, amphibious Animals with the Insects and Reptiles; lastly, the Fish." His taxonomy is based on the usefulness of a "class" of organisms more than its morphology, but this was a common approach at the time and one that reflected the observations of the voyagers from whom his information derived. Dozens of authors are cited, and while many of these are also represented by their "journal and adventures," Green also makes much use of Jean Barbot's account of the West African environment in *Churchill's Collection,* even though Barbot's "adventures" are omitted. Barbot's *Description* was itself a composite text crafted from Barbot's manuscripts and various published works, especially Bosman's *New and Accurate Description of the Coast of Guinea.* Green does not include Bosman's "journal and adventures" either, and he is infrequently cited in the chapters on Sierra Leone, but he appears regularly throughout the rest of the volume.[21]

While the composite nature of the natural history passages lends itself to a variety of voice and content, Green edited them with care. The fact that they were compiled by a single editor who had apparently never visited the region means that the selections are largely made on rhetorical rather than empirical grounds. Green aims for a happy medium, omitting information that deviates from the collective view and paraphrasing descriptions to achieve an even tone. The immediate consequence is that much of the exaggerated language used by Barbot and others is moderated. Despite the alteration of tone, Green footnoted his sources meticulously and alerted the reader to inconsistencies. For example, he includes a lengthy description of "the Birds named *Kubalos,* or Fishers" compiled from reports from six different observers and annotated with eleven separate footnotes. This bird, which is also illustrated by a figure copied, in reverse, from Barbot's account of the Sestro River in *Churchill's Collection* (see figure 4), "is of the size of a Sparrow, with a variegated Plumage [Note: Like a Goldfinch]. The Bill is as long as the whole Body, strong and pointed, and armed, on the Inside, with small Teeth like a Saw. He hovers in the Air, on the Surface of the Water, with so brisk and lively a Motion, as dazzles the Eyes. They abound on both shores, near the Isle of *Ivory,* where they swarm in Millions; their Nests hanging so thick over the Water, that the Negros call them Villages." These nests "hang, by a long Thread, to the Ends of the Smallest Branches that overlook the River; so that at a Distance, they seem like Fruit on the Boughs." The size, plumage, and description of the nest identify this as a species of weaver bird, and the note that "the Negros call [the nest colonies] Villages" suggests the bird is the village weaver (*Ploceus cucullatus*), which is common in the region. The bill and hunting behavior are, however, more characteristic of kingfishers, of which there are several species in West Africa, but kingfishers do not have highly serrated bills, nor do they or village weavers have forked tails, as does the specimen portrayed in Barbot's sketch. Green does his best both to reconcile these contradictory accounts and to moderate the exaggerated language used by the six travelers to describe the birds, but it is clear that doubts remain. Barbot's account in particular is downplayed. Green omits Barbot's hyperbolic assertion that the "ablest artist could not imitate the work of these little creatures" and relegates to a footnote his (unlikely) observation that he saw a thousand such nests in a single tree. The next footnote comments on Barbot's illustration: "The Figure of these Birds and their Nests do not exactly answer this Description; whence it appears, that Barbot, from whom it is taken, has imposed on his Readers." Imposition or not, and although the account of the Kubalo does not quite precisely describe any actual species, Green takes considerable pains to reach an even tone and a plausible description.[22]

Green's moderation of previous accounts does not, however, diminish the overall impression of Africa's biological diversity and plenitude. Although more measured in his language, by synthesizing the work of many others Green is able to include details of a much wider range of organisms. As with his taxonomic scheme, the particulars often emphasize the utility of the plants and animals under discussion (albeit while omitting many of the unsubstantiated claims about the organisms' medicinal efficacy) but this is not always the case. Many wild plants and animals are described purely on the grounds of their appearance and behavior. Stripped of the rhetorical flourishes of the original authors, these descriptions become clear and scientific. Others, however, are assessed in relation to human activities. Gastronomic information is frequently included, as is the utility of stems, leaves, hides, and feathers. Trees are arranged in order of usefulness. "THE most useful as well as common Tree, in these Parts," begins Green, "and indeed throughout all *Africa,* is the Palm-Tree." These are again subdivided by use into "the Date-Tree, the Coco-Tree, the *Areka* Tree, and the Cypress-Palm, or Wine-Tree," and considerably more attention is paid to the products of these trees than to their morphology. Several species of palm later, he notes, "The next useful Tree in this Part of Africa . . . is the Cotton-Tree." An essay on cotton (probably Levant cotton, *Gossypium herbaceum*) follows, and then a paragraph on indigo (*Indigofera tinctoria*) and another on tobacco (*Nicotiana spp.*), the latter both introduced non-native species. Almost half of the section on "roots and plants" is dedicated to the banana, followed by much shorter descriptions of pineapple, watermelon, "manjok," "patatas," "great millet or maez," "small millet," "kuskus," and rice. The taxonomy is not so much a great scale of being as a commercial scale of value. The two chapters on "Beasts, wild and tame" do not follow quite the same pattern. The first commences with the lion (*Panthera leo*) and then works its way through the remaining African big cats. The second chapter, however, begins with the elephant (*Loxodonta spp.*), the source of lucrative ivory, and is then mostly concerned with cattle and antelope, sources of milk, hides, and flesh. Apes, monkeys, and baboons are briefly treated as curiosities but mainly as pests that inflict great economic harm. The chapter on fish emphasizes species that are useful as food, although it also contains much on sharks and a long speculative account of the common torpedo (*Torpedo torpedo*), a fish familiar from antiquity that, at the time of writing, was supposed to have its stunning effect by means of a rapid poison rather than, as is now known, by an electric discharge. The title of the chapter "Birds and Fowl" implicitly divides the class into edible and nonedible birds. The net effect of these chapters is to represent Africa as a site of great biological diversity but also as a place teeming with edible and other useful resources.[23]

A notable feature of *Astley's Collection* is the extent to which it relies on African local knowledge to interpret habitats. This is already apparent in the original narratives, but in Green's synthesis it becomes especially noticeable. By focusing on useful organisms, Green is compelled to show them in use. Thus, the sections on the region's various palms include long accounts of how local people prepare palm oil and palm wine and the manner in which palms are climbed and harvested. The long section on elephants includes much on how they are hunted and eaten by Africans, but also includes detailed accounts of elephant behavior alongside anecdotes of interactions between elephants and local people that were related to European travelers by Africans—several such passages from Bosman's *Description* are included entire. The use of other animals, both wild and domesticated, is also described in detail, including the horses, "of great Beauty and Value," the process by which musk is obtained from civets (*Civettictis civetta*), and the cattle, which "yield a great deal of Milk." In some cases, travelers sought local knowledge to corroborate or refute European mythology. Green notes that "JOBSON says that, by the Report of the Natives on the *Gambra,* there is a Beast in the Country of the Size and Colour of a Fallow-Deer, with a Horn about the Length of a Man's Arm, which, the Author observes, is not like the Unicorn as he is painted; Nor would he probably have heard of such a Beast, if he had not enquired after it. However, *le Maire* informs us, that there are Rhinoceros's here, but never saw any." The simple fact of inquiring after the unicorn requires cultural exchange and linguistic facility but it also demonstrates that Jobson expected, or at least hoped, to receive useful information. These passages imply sustained and careful cross-cultural communication, but they also represent African engagements with organisms and habitats as productive, complex, and rich. The hunting, fishing, and agricultural pursuits they depict emerge from a land of great fertility and diversity and are, most emphatically, the cultural and economic activities of a free people.[24]

Although the reports are compiled from those of slave traders, slavery and the slave trade are barely mentioned in the natural history sections of *Astley's Collection.* There are, however, two exceptions. The second and shorter is in the description of the manatee, of which Green includes another account, having also reproduced Atkins's report of the animal in the story of Tomba's enslavement in the "adventures." In a footnote, Green remarks of the manatee's hide that "they use it in the *West Indies* for Thongs to correct their Slaves, like our Bull's Pizzles." This observation goes no further, although we may note that, although Green sets up an "us and them" dichotomy between British readers and West Indian slaveholders, the difference is merely one of available resources rather than whether whipping is defensible; flogging was a common practice in British civil and military life as well as in the colonies. The longer discussion, a few pages earlier, is in the extended passage on the shark, taken largely from Bosman. In Bosman's original and Green's abridgement, the shark both eats and is eaten. "This Fish," in Green's version, "is the Negros best and most common Food," although found tough by Europeans. To cure them, the African fishermen "lay them seven or eight Days to rot and stink, after which, they are greedily eaten as a Delicacy; and a great Trade is driven in the inland Country with this Commodity." Although the account betrays the author's disgust, the report also offers a scene of free people engaged in fishing, food processing, and long-distance trade. The next paragraph turns the tables completely: "If any Person falls overboard he is a dead Man, unless none of these Fish are near, (which is a Rarity) or he is immediately helped out. When dead Slaves have been cast into the Sea, *Bosman,* not without Horror, has seen the Rapaciousness of these Animals; immediately four or five together shoot to the Bottom, under the Ship, to tear the Corpse to Pieces; at each Bite, an Arm, a Leg, or the Head is snapt-off; and, sometimes, before you can tell twenty, it is every Morsel devoured, Entrails and all." The scene is indeed horrific. Bosman's vivid account, refracted through *Astley's Collection,* contributed to an enduring trope of abolitionist literature in which the shark became a metonym for the violence of the slave trade. Beyond literary texts, sharks did indeed follow slave ships where, in Marcus Rediker's words, "the destruction of corpses by sharks was a public spectacle and part of the degradation of enslavement" in which "the histories of terrorism and zoology intersected." The image of sharks feeding or attempting to feed upon dead or dying captives occurs repeatedly in a wide variety of forms and contexts. The common element is that the reader is invited to compare the slave trader with the shark. Olaudah Equiano (c. 1745–97), for example, writing in 1789, recalled a childhood shipboard experience in which the crew "caught, with a good deal of trouble, a large shark, and got it on board. This gladdened my poor heart exceedingly, as I thought it would serve the people to eat instead of their eating me; but very soon, to my astonishment, they cut off a small part of the tail, and tossed the rest over the side. This renewed my consternation; and I did not know what to think of these white people, though I very much feared they would kill and eat me." In the same year, the abolitionist poet John Jamieson (1759–1838) portrayed an African captive who,

> When from the ring-bolts loos'd to leave the deck,
> Leaps overboard, the partner of his chains,
> Of life less lavish, dragging after him;
> And fills a monstrous shark's deep-forked jaws,
> Expanding to receive its shrinking prey.

Jamieson's now little-remembered verse depicts the horror of an enslaved person jumping to his death. In a better-known poem written two decades later, Percy Shelley (1792–1822) compared the home secretary and foreign secretary with sharks lurking in wait for bodies to be thrown from a slave ship:

> As a shark and dogfish wait
> Under an Atlantic isle
> For the Negroship, whose freight
> Is the theme of their debate,
> Wrinkling their red gills the while:—.

By the 1820s, when the majority of abolitionists were still lobbying for gradual emancipation, the Quaker abolitionist Elizabeth Heyrick (1769–1831) challenged that position by likening slaveholders to criminals and wild animals, asking, "Must it therefore follow, by any inductions of common sense, that emancipation out of the gripe of a robber or an assassin, out of the jaws of a shark or a tiger, must be gradual?" A decade later, the artist J.M.W. Turner (1775–1851) famously portrayed sharks and other sea creatures devouring the bodies of people jettisoned from a slave ship in his painting *Slavers Throwing Overboard the Dead and Dying—Typhoon Coming On* (1840). These are just a few examples among many, but they attest to the enduring horror—and power—of the depiction of captive Africans being thrown to sharks as well as to the profound influence of naturalists' accounts of Africa on abolitionist rhetoric.[25]

Like many of the passages in *Astley's,* the description of sharks originated elsewhere, in this case from Bosman, but was given a wider audience through its inclusion in the collection. Containing hundreds of extracts, thousands of pages, and millions of words, the text is difficult to pin down to a single meaning or interpretation. Green represents the views of both those, like Snelgrave, who justified the slave trade and those, like Atkins and Moore, who found it distasteful. Abolitionists and proslavery apologists alike could find something to support their point of view in *Astley's Collection*—not, one suspects, because Green intended to be politically even-handed but because he sought to deploy the widest possible range of sources. Nevertheless, the Africa represented by Green consists of complex and varied societies located within a diverse and productive environment. At the same time, the slave trade, although part of a complex economic system, is depicted as the cause of numerous injustices. In synthesizing the experiences and views of a great many different observers, in hundreds of iterations and reiterations, he amplified both slavery's cruelty and Africa's productivity in a way that, although perhaps hidden to Green himself, would be highly suggestive to his readers.

MICHEL ADANSON'S VOYAGE TO SENEGAL

Although it had not appeared in either *Churchill's* or *Astley's,* Michel Adanson's *Voyage to Senegal, the Isle of Goree, and the River Gambia,* originally published in Paris in 1757 and translated into English in 1759, is widely cited in abolitionist literature despite the fact that it has almost nothing to say about the slave trade. Adanson (1727–1806) was a French-Scottish botanist who spent five years on a scientific mission to Senegal, established an important system for naming plant families, and dedicated much of his life to creating an encyclopedic account of all known animals, minerals, and vegetables. This was never published, and he died in poverty. The *Voyage to Senegal* was a more modest work of 360 pages in its English translation and, unlike almost every other description of Africa published at this point, it was written by a trained naturalist rather than by a slave trader. This fact was not lost upon the anonymous English translator who, in a preface, notes, "Hitherto we had received but very imperfect accounts from that part of Africa, former adventurers having had no notion of improving their minds, but their fortunes; so that their relations are confined to the *auri sacra fames,* the purchase of slaves, teeth and dust." Although these traders did, in fact, add to European knowledge about Africa, it remains true that dedicated scientific missions of the sort undertaken by Adanson were a new phenomenon in the mid-eighteenth century.[26]

The original French title was *Histoire naturelle du Sénégal. Coquillages* ("Natural History of Senegal. Shells"). It was divided into two parts, the first third the *voyage,* or "journey," and the second two thirds an extended *Histoire des coquillages.* Adanson's typology of seashells proposed a rival scheme of biological classification to that formulated by Linnaeus, making the book one of the first to attempt a systematic analysis of an African habitat in the way that some others had already attempted with the Caribbean. Although Adanson's conchology certainly has the meticulous and highly ordered approach of Hans Sloane's *Voyage to Jamaica,* the book as a whole lacks the breadth of Sloane's *Voyage,* or even that of Griffith Hughes's problematic *Natural History of Barbados.* Its English translator clearly considered the natural history of shells as a niche interest since it is omitted in the English translation, even though it offers the more substantial contribution to science. Instead, the translator presents only the more accessible travelogue that preceded it, renaming the entire book *Voyage to Senegal.*

Adanson arrived in Senegal in early 1749 and settled on the island of Ndar, which he called Senegal Island, now the location of the old town of Saint-Louis. Once the capital of Senegal, this is today a substantial provincial city on the country's arid northwest coast. He traveled frequently to other parts of the region and spent long periods at the French slave-trading fort of Gorée, a small island about three kilometers (two miles) offshore from the current Senegalese capital Dakar. These locations are vividly described in *Voyage to Senegal,* which takes the form of a journal of the author's travels and includes his observations on topography, meteorology, botany, and zoology as well as numerous encounters with local people. The text would more accurately be described as the diary of a naturalist at work than as a work of natural history. As he walks around, armed with a fowling piece, Adanson collects samples of the plant and animal species he encounters by "herborizing and coursing all the way; so that I never returned empty handed. In one place, a plant, or an insect, stopped me; in another, some extraordinary quadruped, or some bird decked with the most beautiful feathers; every object that offered itself to my sight, was new to me." As well as plucking and shooting, he took careful measurements, some using cutting-edge scientific equipment, such as his records of daily temperature taken from the alcohol thermometer and temperature scale invented by his mentor, the entomologist René Antoine Ferchault de Réaumur (1683–1757). Others were less precise, as for example when he used the traditional method of tree hugging to estimate the prodigious girth of the baobab tree (*Adansonia digitata*). The incident is celebrated in the scientific name for the genus. His descriptions of organisms are often clear and precise but are incorporated within the traveler's narrative and not extended, even though many are supported with botanical names in footnotes. Sometimes he withholds information and promises a more specialized publication in future. On a trip up the Senegal River, for example, he notes having discovered "a kind of shrub hitherto unknown to botanists," but he declines to describe it. "I likewise discovered a considerable number of other new plants," he admits, "but it would be no use to mention them here, as I intend to give their description and figure in a particular work."[27]

Adanson shares with other European visitors the habit of emphasizing the prodigality and fecundity of the African environment. This begins with the climate, the heat in particular, but also violent storms, high seas, and a sky "bespangled with stars that shone forth with the brightest lustre." This prodigality is demonstrated by almost every organism he mentions. As with Barbot and others before him, this is not always a positive thing. He is beset with cockroaches and ants, which eat his books and undermine his lodging but, "of all the extraordinary things I observed, nothing struck me more than certain eminences" that turn out to be the homes of termites, which "multiply prodigiously." At night, he is attacked by "great swarms" of mosquitoes that "sucked a good deal of blood, and every morning I had my face disfigured with pimples." On one walk, the sky is darkened by "a cloud of locusts, raised about twenty or thirty fathoms from the ground, and covering an extent of several leagues." These "did all the mischief that can be dreaded from so voracious an insect." Nevertheless, the African environment's ability to recover from such damage is equally remarkable. "One thing which always surprized me," he records, "is the prodigious rapidity, with which the sap of trees repairs any loss they may happen to sustain in that country: and I was never more astonished, than when, upon landing four days after that terrible invasion of locusts, I saw the trees covered with new leaves." It is not just insects, however, that wreak destruction. "The most terrible scourge of all was a large species of yellow and black sparrows, clouds of which fell like hail upon the grain; and when they had spread desolation in one quarter, they flew to another."[28]

Not all insects are pests. "One day," says Adanson, "I saw the roof of a house, the surface of which was sixteen square feet, covered with a lay or bed, four fingers thick of bees." Not only is this "evident proof of the prodigious number of those insects," but he also finds the honey "infinitely superior, both in delicacy and taste, to the best honey collected in the southern parts of France." Other organisms are equally marvelous. The Senegal River contains "plenty of excellent fish, as thornbacks, soles, monstrous large rock fish and a great many *tree-oysters,*" which cling to the trunks of mangroves. Further afield, "the rocks with which the island of Goree is surrounded produce an infinite number of shell and other fish," while in the village of Ben (probably Hann, now a neighborhood of Dakar) a "ship's crew filled their boat till it was ready to sink, leaving the rest on the sea-shore. In any other country, such a capture of fish would, without all doubt, pass for a miracle." On the low-lying coastal islands north of Saint-Louis, now part of neighboring Mauritania, Adanson found a marsh that "abounds with aquatic birds, such as curlieus, wood-cocks, teals, and wild ducks." A hunter as well as a naturalist, Adanson noted that the latter "shew themselves by thousands, and you kill them as it were by thousands. It is not uncommon to see thirty of them drop at one shot." Such fecundity arises from the soil, which is "rich and deep, and amazingly fertile: it produces spontaneously and almost without cultivation, all the necessaries of life." It is "so very fruitful, that a great many plants yield several times a year." Husbandry is also greatly facilitated. "There is not perhaps a country in the world where poultry are more common. They breed turkeys, Guinea hens, geese, ducks, and a prodigious number of fowls. Their pigeons are in admirable perfection; and their hogs multiply very fast." Wherever one turns, the African environment gives rise to prodigies of size, number, or magnificence.[29]

Adanson's account of African people is less enthusiastic than his description of the land, but he was equally keen to investigate and indeed dedicated the first months of his stay to learning to speak Wolof. He shared the prejudices of many Europeans, remarking on "the laziness and indolence of the Negroes" on several occasions throughout the book and remarking that poor Africans, at least, were "the most artful beggars, and the most dexterous thieves, in the universe." His account of Senegal's ruling families is somewhat more nuanced. After visiting and dining with Baba-sec, "master of the village of Sor" (today a mainland suburb of Saint-Louis), he begins to conceive of Senegalese society in a more positive light, albeit refracted through European mythology and ideas of African primitivism. "Which way soever I turned my eyes on this pleasant spot," he writes, "I beheld a perfect image of pure nature: an agreeable solitude, bounded on every side by a charming landskip; the rural situation of cottages in the midst of trees; the ease and indolence of the Negroes reclined under the shade of their spreading foliage; the simplicity of their dress and manners; the whole revived in my mind, the idea of our first parents, and I seemed to contemplate the world in its primeval state." Eighteenth-century readers would have recognized this Edenic scene both as decorously metaphorical and as a reference to the philosophical debate over the essential moral nature of the human species. While few, if any, Enlightenment thinkers believed that human beings had ever literally existed in a "state of nature" anywhere in either space or time, imagining such a condition was a useful thought experiment, allowing philosophers to probe the nature of humanity itself. Adanson's vision of Senegal, while demoting Africans to a state of relative primitivism, asserts the fundamental goodness of human nature, and it is significant that he reaches this view after having shared couscous on relatively equal terms with a local family after having spent some months learning how to communicate in their language. "I was not a little pleased with this my first reception at the lord of Sors," he concludes. "It convinced me, that there ought to be considerable abatements made, in the accounts I had read and heard every where, of the savage character of the Africans; and I was of opinion, that this should not extend to the natives of Senegal." In Adanson's case, familiarity breeds respect and he begins to question his assumptions. After two months, he notes that he "had observed great humanity and sociableness" among both "the Moors and the Negroes." In a moment of self-awareness, after an "extraordinary and unexpected scene" in which a group of children approach him with much curiosity, "it came into my head, that my colour, so opposite to the blackness of the Africans, was the first thing that struck the children." Adanson realizes that he is the exception here and that his color, although a marker of difference, is not a mark of superiority. While the comments about idleness and thievery cast a shadow over his narrative, his account of African society is otherwise generally positive.[30]

At both Saint-Louis and Gorée, Adanson must have come into close contact with slave traders, but he passes over these encounters in silence. The only reference to slave trading appears in the English edition, in a footnote supplied by the translator. Adanson does speak frequently of "my negro" or "my negroes," but there is no evidence that these individuals are enslaved and it is possible that they were paid local guides and servants. The reason for his silence is unclear. He may simply have decided that the slave trade was beyond the scope of a naturalist but, given his largely favorable account of African society, he may have disapproved of the trade but did not wish to offend the slave traders who were his hosts in Senegal. What we can say with certainty is that Adanson's account of the region represents a fertile and diverse environment, home to complex and varied societies. Adanson's assessment of Africa as brimming over with natural resources was very much shared with others who depicted the continent. Unlike many, however, Adanson did not represent the people of Africa as an exportable commodity, but it is noteworthy that even those who did rarely presented the slave trade as an unalloyed good. Slavery in their accounts existed as part of the natural landscape of African society, as it did in many parts of the world, but the Atlantic slave trade driven by European demand is repeatedly shown in these narratives to modify, increase, or corrupt indigenous practices that had arisen organically and were rooted in custom. Abolitionist readers of Churchill, Astley, or Adanson would find much to confirm that the trade was the source of great suffering, but they would also learn that Africa had much more to offer than simply slaves, teeth, and dust.[31]

PART TWO

Deploying the Archive

CHAPTER 5

"*The groans, the dying groans, of this deeply afflicted and oppressed people*"

ANTHONY BENEZET AND THE NATURAL HISTORY OF ATLANTIC SLAVERY

By the third quarter of the eighteenth century, naturalists, agriculturalists, and scientifically informed travelers working in the service of slave traders and the plantocracy had built up a large body of literature concerned with the nature and culture of both the African and the Caribbean environments. Despite some anxieties expressed by seventeenth-century observers such as Richard Ligon, most of this literature either passively accepted plantation slavery as a natural phenomenon, albeit one artificially introduced and then naturalized to the Americas, or actively advocated for its continuation, and with it the slave trade that maintained it. The attitudes expressed in these writings were no doubt depressing reading for the small but growing band of antislavery campaigners, at first mainly Quakers and Methodists, who were increasingly active from the 1750s onward. The information these works contained, however, was invaluable. Agricultural manuals and accounts of the natural history of Africa and the Caribbean included copious information about slavery and the slave trade, proving to be a fruitful and compendious archive from which authors writing in support of the nascent abolition movement could draw. The first wave of abolitionists mined this corpus assiduously, using the evidence accumulated in environmental literature to bring the cruelty of slavery into the public gaze. At the same time, they challenged, undermined, and ultimately inverted the assumption of colonial naturalists that slavery was an inevitable part of the natural order, repositioning slavery and the slave trade as unnatural.

This chapter, the briefest of this book, is nevertheless its fulcrum. It shows how the most important abolitionist writer of the mid-eighteenth century, Anthony Benezet (1713–84), raided this archive, transformed it, and repackaged it in an easily digestible form that profoundly influenced the coming generation of abolitionists. Benezet's reading interests, and the books that by chance he had available to him, would therefore have a disproportionate influence on the character of the British abolition movement. He published eight antislavery texts, of which his *Some Historical Account of Guinea*, published in Philadelphia in 1771 and reprinted verbatim in London the following year, was both the most substantial and the most influential, inspiring a generation of abolitionist writers on both sides of the Atlantic and arguably providing the spark that ignited widespread but largely quiescent antislavery sentiment into abolitionist activism. The book draws extensively on natural history writing but also adopts the encyclopedic method of its principal source texts, *Astley's Collection* in particular. Scholars generally agree that Benezet's method was effective because he drew on the travel literature of supposedly unbiased slave traders, especially those reproduced in Churchill's and Astley's collections. Benezet not only draws from these authors' descriptions of people and societies in their topographies and ethnographies but also borrows from their expertise and authority in natural history, a field of scientific inquiry that transcends even as it serves their economic interests. *Some Historical Account* can therefore be understood as a natural history of Africa as well as an account of its people and principal locations. At the same time, and despite its title, much of *Some Historical Account* is actually concerned with the Caribbean, based largely on accounts of naturalists' visits to the islands. *Some Historical Account of Guinea* is, therefore, best understood as a natural history of the Atlantic slave trade, in which slavery is represented as an unnatural perversion of the divinely ordained natural order.

Named Antoine at birth, Benezet was born in St. Quentin, northern France, to a family of Huguenots who fled to London in 1715 before settling in Philadelphia, where they joined the Quakers. His early attempts at a career in trade unsuccessful, in 1739 he started as a schoolteacher in Germantown, later joining the famous Friends' English School of Philadelphia, where he was noted for being a fine teacher and for his dislike of the severe discipline then common. In 1750, he set up an evening class for Black children in his own home. Four years later, he inaugurated the first public girls' school in America and, in 1770, with the support of the Society of Friends, he founded the Negro School in Philadelphia. He taught at both schools until his death in May 1784. From at least the early 1750s, Benezet had become a firm opponent of slavery, combining with John Woolman (1720–72) and others to persuade Pennsylvanian Quakers to prohibit slave trading within their community, and writing antislavery tracts and pamphlets that he published at his own expense. This marked the culmination of a process that had been underway for some time. From the late seventeenth century onward, Quakers engaged in a long and sometimes highly charged discussion about whether slave trading or slaveholding were consistent with their values and permissible within their communities, a debate that began in Barbados but was particularly intense in the North American colony of Pennsylvania. Eventually in 1761, following a similar decision made three years earlier in Pennsylvania, the London Yearly Meeting, the Quakers' weightiest grouping, prohibited Friends from slave trading on pain of disunion. From that point onward, the Society of Friends effectively became the world's first antislavery organization and Quakers were often at the center of national and local abolition societies as they arose during the 1780s and '90s. Benezet was one of the key instigators of that process.[1]

SOME HISTORICAL ACCOUNT OF GUINEA

Benezet is today less widely known than Woolman, with whom he worked closely, and he has not captured public attention in the way that some later publicly and politically engaged abolitionists such as William Wilberforce (1759–1833) or William Lloyd Garrison (1805–79) have done. Nevertheless, historians of antislavery generally agree on his pivotal early role in galvanizing support for abolitionism, even though the importance of that role is more often asserted than examined. As a Quaker who rarely traveled far from Philadelphia, he understood that his contribution would need to be textual if it were to have lasting impact. Between 1754 and 1783 he therefore published no fewer than seven antislavery tracts, plus an augmented edition of John Wesley's 1774 *Thoughts upon Slavery,* which had itself included lengthy extracts from Benezet's own work. The centerpiece of this literary output was *Some Historical Account* which, argues David Crosby, "became a kind of bible for later abolitionists in England and America," and "also became the standard early-nineteenth-century source for African history, with parts of it incorporated whole into early editions of the *Encyclopedia Britannica.*" Jonathan Sassi notes that "the path that he blazed was soon followed by a host of other antislavery figures, which, in turn, called forth the renewed efforts of proslavery writers," while Christopher Brown comments that the book would inspire a generation of abolitionist writers and activists such that "the first British writings to condemn the slave trade at length relied heavily on the information Benezet compiled and the generic conventions he pioneered." On the other hand, Eric Herschthal notes the book's influence but argues that "one of the problems with Benezet's *Historical Account* was that the natural histories it cited were decades old. . . . Benezet had no scientific standing, and he stacked his book with out-of-date observations." This is undeniably true, but it may not have been especially relevant to the book's success as abolitionist literature. The evidence Benezet amassed was compendious, apparently impartial, and it carried the weight of generations. While it may not have satisfied scientists at the cutting edge, it was precisely the right mix to inspire the new generation of abolitionists.[2]

The influence of *Some Historical Account* may be undisputed, but only a handful of scholars have offered detailed analysis of the writing on which Benezet's reputation rests. Those few who have done so agree on certain aspects of his work. All note that Benezet made extensive use of African travel narratives, in particular the Churchill and Astley collections. Most cite from Benezet's own methodological statement in the opening chapter of *Some Historical Account* where he explains that he "will here give some account of the several divisions of those parts of Africa from whence the Negroes are brought, with a summary of their produce; the disposition of their respective inhabitants; their improvements, &c. &c., extracted from authors of credit; mostly such as have been principal officers in the English, French and Dutch factories, and who resided many years in those countries." The decision to use sources closely allied to the slave trade, all agree, was Benezet's masterstroke. "Using the travelers' narratives," says Benezet's biographer Maurice Jackson, "was a brilliant rhetorical strategy on his part: no one could accuse the authors of a pro-African bias or of any animosity toward slavery." Indeed, "by appealing to the authority of the African travel narratives," argues Sassi, "Anthony Benezet cast a paradigm that would influence writers on slavery, for and against, white as well as black, well into the nineteenth century." Sassi builds on a long-standing perception, expressed by C. Duncan Rice and others, that Benezet's choice of evidence represented African societies as naïve. "Benezet sampled from the travel literature only selectively," Sassi argues, and "did not build the strongest case available to him for the full human equality of African people. Instead, he confined himself to depicting the Africans' 'innocent simplicity.' " Indeed, he contends, "there was a great deal of material contained in the travel accounts that flatly contradicted Benezet's portrayal." Jackson, by contrast, argues that "far from presenting a naïve utopian view of Africa, as C. Duncan Rice and others have claimed, Benezet employed a careful calculus to make his selections," each of which offered evidence to support one of "four premises" grounded in African realities rather than utopian vision. These premises, set out at the start of the opening chapter of *Some Historical Account,* can be summarized simply: Africa is rich in natural resources, its inhabitants are fully human and capable, African forms of governance and commerce are "commendable," and Africans "might have lived happy if not disturbed by the Europeans." While these premises do not always map precisely onto the chapter structure of the book, they do accurately reflect the book's major themes and arguments about Africa.[3]

While the interpretations put forward by Jackson, Sassi, and others are important in building up a picture of Benezet's method and choice of content, taking note both of what he included and what he omitted, they tend to overlook Benezet's interest in the American colonies in addition to Africa, the extent to which his sources are natural histories as well as travelers' tales, and the fact that his main sources were in general highly edited anthologies rather than unmediated primary texts. Indeed, I suggest, Benezet's reliance on *Astley's Collection* explains much of the form and structure of *Some Historical Account.* As we saw in chapter 4, rather than reproducing travelers' accounts of Africa verbatim, the collection's editor John Green separated them out into "journal and adventures" and "remarks on countries." Under the former category he reproduced specific incidents in the travels of individuals, whereas under the latter he collated the ethnographic and biological observations of the travelers, combining them, in effect, into unified natural histories of the various African regions. In *Some Historical Account,* Benezet follows Green's lead. Not only is *Astley's Collection* Benezet's most important source, it also provides the method. Benezet selects his material to support his key premises, as one would expect, but he also synthesizes and paraphrases many texts, supporting them with copious footnotes, as Green had done, to produce a composite account of African environments and societies that in form closely resembles Green's encyclopedic "remarks" in *Astley's Collection.* In this sense we can see *Some Historical Account* as part of a continuum of reference works that begins with Ephraim Chambers's *Cyclopædia,* on which Green was employed, continues with *Astley's Collection* and *Some Historical Account,* and culminates in large sections of Benezet's work being incorporated into the *Encyclopedia Britannica.* Adopting Green's form was undoubtedly an important contribution to the lasting and widespread impact of Benezet's work. It is doubly significant since Green's composite "remarks on countries," in his own words, add up to "*a System of* Modern Geography and History, *as well as a Body of Voyages and Travels.*" Both Green's and Benezet's readers would have understood the word *history* to mean natural as well as political history, and *Some Historical Account* indeed includes substantial material in both categories. Several chapters recount the chronological history of European encounters with Africa, but these, although centrally positioned, comprise only about a fifth of the text and, despite the title, *Some Historical Account of Guinea* contains accounts of America and the Caribbean as well as Africa. The book as whole offers, therefore, both a description and a critique of slavery and the slave trade as it operated across the entire Atlantic world in Benezet's time and before.[4]

Benezet begins with physical description of the African continent. The first three chapters set out to demonstrate Africa's great abundance of natural resources, depicting, as Green and others had done before, a continent that was home to many complex and varied societies rooted in a highly diverse and productive environment. Soil is a major preoccupation in the opening chapters of the book, both literally and figuratively underpinning African agriculture and society. The first mention is drawn from Michel Adanson's 1759 *Voyage to Senegal*, which was not included in *Astley's Collection* but which, as we have seen, shares its conception of Africa as a place of prodigious fertility. Adanson describes the soil, in Benezet's quotation, as "rich and deep, and amazingly fertile; it produces spontaneously, and almost without cultivation, all the necessaries of life, grain, fruit, herbs, and roots. Every thing matures to perfection, and is excellent in its kind." Benezet abridges his quotation from Adanson in much the same way as Green abridged his sources, and both the method and the sentiment are reproduced in Benezet's quotations from other visitors to Africa. Benezet quotes Jean Barbot, as he was represented in *Churchill's Collection*, as remarking that "the soil is very good and fruitful in corn and other produce, which it affords in such plenty, that besides what serves for their own use, they always export great quantities for sale." He cites William Snelgrave, as abridged by Green, who thought that "the country appears full of towns and villages; and being a rich soil, and well cultivated, looks like an entire garden." In some places Benezet adopts Green's encyclopedic method entirely. Toward the end of chapter 3, for example, Benezet notes, "In the Collection it is said, that both in Kongo and Angola, the soil is in general fruitful, producing great plenty of grain, Indian corn, and such quantities of rice, that it hardly bears any price, with fruits, roots, and palm oil in plenty." At this point, Benezet is abridging, collating, and frequently paraphrasing material found in the "remarks" sections of *Astley's Collection,* material that was itself already composed of abridgements and collations. This is a recurring feature. The words "in the Collection" appear a dozen times in *Some Historical Account,* in almost every case indicating that Benezet is using Green's method as well as his content. At other points, Benezet dispenses with the formality of mentioning "the Collection." Chapter 2 begins in his own words describing "that part of Guinea known by the name of the *Grain,* and *Ivory Coast.*" In this region, says Benezet, clearly synthesizing many of Green's passages, "the soil appears by account, to be in general, fertile, producing abundance of rice and roots; indigo and cotton thrive without cultivation, and tobacco would be excellent if carefully manufactured; they have fish in plenty; their stocks greatly increase; and their trees are loaded with fruit. They make a cotton cloth, which sells well on the coast. In a word, the country is rich, and the commerce advantageous, and might be greatly augmented by such as would cultivate the friendship of the natives." Chapter 3 begins in a similar vein, describing "the kingdom of Benin, which, though it extends but about 170 miles on the sea, yet spreads so far inland, as to be esteemed the most potent kingdom in Guinea. By accounts, the soil and produce appear to be in a great measure like those before described." In both cases, the character of the nation proceeds literally from the quality of its soil. The soil is rich and productive and the commercial opportunities for Europeans prepared to "cultivate the friendship of the natives" extend considerably beyond gold, ivory, and slaves. In both passages, Benezet imports Green's encyclopedic method and style as well as selecting from and paraphrasing the contents of his *Collection.*[5]

Benezet is not interested only in soil, of course. The opening chapters of *Some Historical Account* stack up description after description of both natural and agricultural fecundity, prodigality, and plentitude. Such natural riches support societies that are, or would be without European interference, happy, wealthy, and wise. In a passage that has attracted particular critical attention, Benezet cites at length, albeit with some paraphrasing, from sections of Adanson's *Voyage to Senegal* in which Adanson presents West Africa as a fecund and self-restoring paradise where "with little labour and care, there is no fruit nor grain but grow in great plenty" and where he was never "more astonished, than when landing four days after the locusts had devoured all the fruits and leaves, and even the buds of the trees, to find the trees covered with new leaves." The long quotation concludes with Adanson's vision of Senegal as a natural paradise that had "revived in my mind the idea of our first parents and I seemed to contemplate the world in its primitive state." As we saw in chapter 4, this passage, while relegating Africans to a "primitive state," asserts the fundamental goodness of human nature, which may be ravaged by the locusts of war and avarice, but which would always grow back like the leaves of the African trees he so admired. In Benezet's hands, this tableau becomes powerful evidence for the innate virtue of Africans, African society, and the African environment, and stands as an image against which the despoliations of the slave trade could be measured. We should note, however, that this reading emerges from Benezet's strategic quotation from Adanson and not directly from Adanson himself. In the original text, the passage about the locusts had appeared a hundred pages after the Edenic vision. Much like Green in the "remarks" sections of *Astley's Collection,* Benezet is extracting from and conflating widely separated passages to create a composite image.[6]

Although derived from natural histories that in most cases paid little or no attention to the conventions of natural theology, this compound image was refracted through the lens of Benezet's faith. Jackson notes that Benezet's impression of African society, here and throughout the book, "resembled Quaker notions of a 'peaceable kingdom.' " Sassi agrees, arguing that "Benezet presented West African peoples as proto-Quakers in their virtues of industry, honesty, and integrity, their social ethic and peacefulness, and their promising, if still incomplete, religious beliefs." Inhabitants of a highly productive environment, "Africans enjoyed their continent's abundance in peace and simplicity, and they exhibited numerous other Quaker virtues such as reverence, charity, and industry. Only European slave traders had spoiled this picture of innocence." Benezet again quotes both generously and strategically to establish how this spoiling takes place. He depicts considerable human misery, but this is buttressed by a broader environmental case against the slave trade in which he maps the effects of the trade onto African landscapes, showing that it is the cause of environmental despoliation as well as human misery. The direct cause of this devastation is the warfare the slave trade produces. This is an important argument for a pacifist Quaker, and his argument that the slave trade has been the cause of devastating wars in Africa is a fundamental principle of the book.

> Tho' wars, arising from the common depravity of human nature, have happened, as well among the Negroes as other nations, and the weak sometimes been made captives to the strong; yet nothing appears, in

the various relations of the intercourse and trade for a long time carried on by the Europeans on that coast, which would induce us to believe, that there is any real foundation for that argument, so commonly advanced in vindication of that trade, viz. "*That the slavery of the Negroes took its rise from a desire, in the purchasers, to save the lives of such of them as were taken captives in war, who would otherwise have been sacrificed to the implacable revenge of their conquerors.*" A plea which when compared with the history of those times, will appear to be destitute of Truth; and to have been advanced, and urged, principally by such as were concerned in reaping the gain of this infamous traffic, as a palliation of that, against which their own reason and conscience must have raised fearful objections.

Benezet augments this position in chapter 5 by arguing that "the Europeans are the principal cause of these devastations" and that "it was long after the Portuguese had made a practice of violently forcing the natives of Africa into slavery that we read of the different Negro nations making war upon each other and selling their captives." He quotes from the slave trader William Smith who, in his 1744 *A New Voyage to Guinea,* had reported that another trader told him that "we Christians introduced the traffic of slaves; and that before our coming they lived in peace." Benezet unambiguously lays the blame for warfare in Africa at the door of European nations, considerably undermining thereby the argument often made by apologists for the slave trade that enslavement of prisoners of war was customary and lawful in African legal systems. He moves beyond his understanding of African positive law, however, to address the relationship between warfare and slavery in natural law. Wars do arise as a consequence of human nature, he argues, and Africans are no different in that respect from any other group of human beings. Around the world, and throughout history, defeated nations have been taken into captivity. As a Quaker, Benezet recoiled against such incidents, but he accepts their universality in the context of original sin, which he calls "the common depravity of human nature." In other words, warfare and captivity in defeat are harsh but natural aspects of the human condition in Africa as elsewhere. The European trade in slaves, however, has no basis in this common experience, while warfare itself is unusual in African societies in their natural state. Benezet builds on the perception of those travelers, reproduced in *Astley's Collection* and elsewhere, who saw the trade as a corruption of indigenous practices that had arisen organically and were rooted in custom, but he strengthens those accounts to show that the European demand for slaves was not merely a corruption or an intensification of what was naturally occurring, but was in fact an unnatural practice, contrary both to natural law and to his faith.[7]

Benezet's depiction of the slave trade as unnatural has another facet. While warfare has a direct human cost in deaths, injuries, and enslavement, its indirect costs often occur as environmental damage. In *Some Historical Account,* this is often measured by population decline, in part as a commentary on the capture of the inhabitants but also as a reflection of the harm done both to agriculture and to natural habitats. In support of this observation, Benezet paraphrases extensively from Willem Bosman who, in Benezet's abridgment, praises the country of Ante (probably Bosman's rendering of the kingdom of Ashanti in present-day Ghana) as "a land that yields its manurers as plentiful a crop as they can wish, with great quantities of palm wine and oil, besides being well furnished with all sorts of tame, as well as wild beasts; but that the last fatal wars had reduced it to a miserable condition, and stripped it of most of its inhabitants." Likewise, says Bosman, "the adjoining country of Fetu" was "formerly so powerful and populous, that it struck terror into all the neighbouring nations; but it is at present so drained by continual wars, that it is entirely ruined; there does not remain inhabitants sufficient to till the country." Taken together with the earlier commentary on warfare, the point could not have been made more clearly. The slave trade is a cause of wars in Africa, and warfare despoils what would otherwise have been a fruitful environment for human farmers.[8]

An analogous argument emerges from Benezet's discussion of African river systems. Rivers were the main highways for Europeans in both Africa and America in the eighteenth century and so it was inevitable that Benezet's knowledge of the African interior would primarily have derived from accounts of African rivers. Thus, the opening chapter of the book contains "a general account of Guinea; particularly those parts on the rivers Senegal and Gambia," and he continues to describe the coast in terms of its river systems to as far south as Luanda, the principal city and now capital of Angola. His discussion reveals both how far inland the slave trade extends and how damaging it is to the African environment. "The great river Senegal," he tells us, "is said to be navigable more than a thousand miles." He selectively quotes from Green's translation, offered in volume 2 of *Astley's Collection,* of the memoirs of French slave trader André Brue (1654–1738), themselves drawn from a composite text edited by Jean-Baptiste Labat (1663–1738) and including much extraneous material. In Benezet's version, Brue had attested, "The farther you go from the sea, the country on the river seems the more fruitful and well improved; abounding with Indian corn, pulse, fruit, etc. Here are vast meadows, which feed large herds of great and small cattle, and poultry numerous: The villages that lie thick on the river, show the country is well peopled." Likewise, the Gambia River, according to Benezet, "is navigable about six hundred miles up the country." He again turns to, abridges, and paraphrases Green's translation, abridgement, and paraphrase of Labat's edition of Brue's memoirs. As Benezet renders it, Brue "was surprised to see the land so well cultivated; scarce a spot lay unimproved; the low lands, divided by small canals, were all sowed with rice, etc., the higher ground planted with millet, Indian corn, and peas of different sorts; their beef excellent; poultry plenty, and very cheap, as well as all other necessaries of life." This is again a depiction of African plenitude, but we should note that the relevant passages in *Astley's Collection,* marked with the side note "Great Fertility," include both nature notes and botanical illustrations and Green himself discusses, in his introduction to Brue, his admittedly somewhat haphazard but nonetheless interesting natural history. Benezet's deployment of Brue's natural history serves a more strategic purpose. He wants his readers to take home the clear message that African rivers are more populous and better cultivated the further inland one travels. It is at the coast, he implies, where Africans and Europeans do business, that the environmental damage and depopulation brought about by the slave trade is at its most severe. Benezet's argument relies on strategic quotation. The slave traders who cause that damage are at the same time the most reliable witnesses to its shocking effects, but only while they are viewing their environment through the scientific lens of natural history rather than in the sights of commercial exploitation.[9]

SOME HISTORICAL ACCOUNT OF THE CARIBBEAN

Benezet recognized that this exploitation was a transatlantic business. While the reference to Guinea in the book's title implies that it is exclusively concerned with West Africa, in fact, two of its fifteen chapters are concerned mainly with the Middle Passage while four focus on Britain's American colonies, particularly in the Caribbean. The chapters on the Middle Passage are largely drawn from the same sources as those concerning the African continent—in most cases, explicitly from the versions in *Churchill's Collection* and *Astley's Collection.* Since many of the accounts in the collections tended to elide or downplay the cruelties of the transatlantic passage, Benezet also inserted the testimony of an anonymous slave trader who witnessed both cruelty and an insurrection and shared his experiences in the hope that he might "contribute all in my power towards the good of mankind, by inspiring any individuals with a suitable abhorrence of that detestable practice of trading in our fellow-creatures, and in some measure atone for my neglect of duty as a Christian, in engaging in that wicked traffic." John Coffey has made a convincing case that Benezet's anonymous author was John Newton (1725–1807), who would later become an evangelical cleric and prominent abolitionist. If that is the case, it points to Benezet's increasing efforts to build an international antislavery network as well as to seek out new evidence, but the physical distance between Newton in England and Benezet in Pennsylvania also reminds us that Benezet was proceeding with some caution. As a resident of a British colony in North America, he was a daily witness to the enslavement of people brought forcibly from Africa as well as of their children and grandchildren who had known no other home but America. He was committed to improving the condition of enslaved Pennsylvanians, not only through his efforts to stamp out slaveholding among the colony's Quaker population but also through his program of educating enslaved children. In both endeavors he ran a substantial risk of antagonizing his fellow colonists, which may in part explain why the critique of colonial slavery in *Some Historical Account* focuses more on Caribbean plantation slavery than on enslavement closer to home. On a more textual level, Benezet was forced to adapt the way he gathered and presented his evidence since America was not well represented in either Churchill's or Astley's collections. The Churchills did include a few American narratives, mostly from the early seventeenth century, but Green includes very few, apparently for no better reason than that he fell out with his publisher before he could complete the planned American volume. Benezet nevertheless retains Green's method of extraction and abridgement in his discussions of the Caribbean, but he selects his primary sources himself. Most of these are natural histories.[10]

Benezet turns to the Caribbean in chapter 8 of *Some Historical Account,* a chapter that arguably forms the moral and structural centerpiece of the book. It rests on the evidence of four authors, three of whom were naturalists or natural history writers. Benezet begins his account of the condition of enslaved people in the British colonies with evidence from Griffith Hughes. Benezet does not comment on Hughes's eccentric botanical method, but he was certainly convinced by the demographic arguments that Hughes had offered in the introduction to his *Natural History of Barbados.* Paraphrasing Hughes, he notes, "We are told by Griffith Hughes" that "there were between sixty-five and seventy thousand Negroes, at that time, in the island, tho' formerly they had a greater number: That in order to keep up a necessary number, they were obliged to have a yearly supply from Africa: that the hard labour, and often want of necessaries, which these unhappy creatures are obliged to undergo, destroy a greater number than are bred there." The paraphrase is mostly due to Benezet changing Hughes's "we" to the more detached "they." Hughes sees no irony in claiming, in his own words, that "we are obliged, in order to keep up a necessary Number, to have a yearly Supply from *Africa,*" but Benezet recognizes neither obligation nor necessity and clearly marks the passage as Hughes's assertion. The careful deployment of Hughes's words is important. The argument that the slave trade would be unnecessary were planters to treat their slaves with sufficient humanity that they would naturally increase in number would become a central plank of abolitionist discourse before 1807, when abolitionists were largely seeking an end to the slave trade but not yet to slavery itself—many believed that abolition of the former would inevitably lead to the withering away of the latter. Although Hughes himself saw the constant importation of captive Africans as necessary to business as usual rather than as a shocking indictment of plantation society, it is nonetheless appropriate for Benezet to cite a naturalist to emphasize this important demographic point. It is also consistent with his strategy of turning the words of slave traders and proslavery apologists against them.[11]

Hughes's demographic evidence is followed up by a similar account from the only wholly political history cited in the chapter, William Burke's (c. 1730–98) 1757 *Account of the European Settlements in America,* which he may have coauthored with his celebrated relative Edmund (1729—97). Benezet then almost immediately turns to Thomas Jefferys's 1760 *Natural and Civil History of the French Dominions in North and South America.* Jefferys (c. 1719–71), the king's cartographer, was better known for mapmaking than natural history, but *Natural and Civil History* resembles *Astley's Collection* in its encyclopedic method and use of extracts from multiple sources. Although it is not the work of a naturalist, it presents more than enough natural history to qualify it as natural history writing. It also had considerable authority conferred upon it by its association with the king. Jefferys arranges his material by region and includes eight chapters on "the Natural History of the Antilles." These begin with climate and soil, proceeding to plants, quadrupeds, birds, and fish, before concluding with two ethnographic chapters on "the original inhabitants" and "the negroe slaves of the Antilles." Benezet extracts passages from the last of these in which Jefferys claims, "It is not our intention, in this place, to consider whether one species of mankind has a right to enslave another." This claim is undercut by the following passage, from which Benezet quotes at length, in which Jefferys argues that "it is impossible for a humane heart to reflect upon the servitude of these dregs of mankind, without in some measure feeling for their miseries, which end but with their lives." Benezet changes the word "humane" to "human," possibly in error although perhaps to counterpoint the following line in which Jefferys notes that enslaved people in the Antilles "are in a measure reduced to the condition of beasts of burden." Benezet omits, however, Jefferys's satirical contrast between the enslaved laborers who, "in the midst of all these hardships . . . enjoy an almost uninterrupted state of good health, while their masters, glutted with the conveniences and pleasures of life, are subject to an infinite number of disorders." Benezet no doubt lamented the consequences of idleness and overconsumption that afflicted many planters just as much as Jefferys, but at the same time he must have known perfectly well from personal experience that enslaved people often fell ill, and frequently died, from a multitude of avoidable diseases and injuries, many deliberately inflicted upon them. Satire will not do if it misrepresents the facts. The sentimental tone that Jefferys adopts is, however, acceptable to Benezet, who frequently deploys the then fashionable rhetoric of sensibility to draw attention to plantation suffering.[12]

In the second half of this important eighth chapter, Benezet turns to one of the eighteenth century's most celebrated naturalists, quoting extensively but selectively from the distressing account of punishments and tortures Hans Sloane had presented in his 1707 *Voyage to Jamaica.* The passage begins by citing Sloane explicitly if not precisely:

> Sir Hans Sloane, in the introduction to his natural history of Jamaica, in the account he gives of the treatment the Negroes met with there, speaking of the punishments inflicted on them, says, page 56. "For rebellion, the punishment is burning them, by nailing them down to the ground with crooked sticks on every limb, and then applying the fire, by degrees, from the feet and hands, burning them gradually up to the head, whereby *their pains are extravagant.* For crimes of a less nature, gelding or chopping off half the foot with an axe.—For negligence, they are usually whipped by the overseers with lance-wood switches.—After they are whipped till they are raw, some put on their

skins pepper and salt, to make them smart; at other times, their masters will drop melted wax on their skins, and use several *very exquisite torments.*"

As with the extracts from Hughes, this is both a paraphrase and an abridgement, and Benezet also adds emphasis, with the effect of concentrating and intensifying the emotional impact of the already harrowing passage. Sloane is a substantial witness and Benezet leans considerably on the weight of his authority, which is derived partly from his posthumous celebrity as president of both the Royal Society and the Royal College of Physicians, but also, in this passage, from the fact that the testimony is from a noted scientist writing in a book that aimed at accuracy and precision.[13]

Sloane appears at first glance to be the principal source for the remaining ten paragraphs of the chapter since Benezet names no further names. In fact, beyond this extract, most of what follows seems to have been of Benezet's own composition, much of it revising material that had appeared in 1767 in his pamphlet *A Caution and Warning to Great Britain and Her Colonies.* Like *Some Historical Account,* this had been written with the clear intention of changing minds on both sides of the Atlantic. Benezet had sent it at his own cost to many influential people, but it had not made much impact, perhaps because of the fulsomeness of its rhetoric. *Some Historical Account* has in general a quieter style, with more frequent references and citations, which tends to create a stronger sense of authority. Chapter 8 is something of an exception to this rule. Although Benezet begins by quoting strategically from the natural history writing of Hughes, Burke, Jefferys, and Sloane, having once established the moral and scientific weight of the chapter on their authority, he then, and without indicating the transition, offers an original composition that builds on the emotional tone set by the earlier authors. The intention seems to be to persuade readers to accept statistical evidence on emotional terms: another example of Benezet's sentimental rhetoric. At one moment, Benezet coolly notes of slaveholders that it is "amazing to hear these men calmly making calculations about the strength and lives of their fellow men. In Jamaica, if six in ten of the new imported Negroes survive the seasoning, it is looked upon as a gaining purchase. And in most of the other plantations, if the Negroes live eight or nine years, their labor is reckoned a sufficient compensation for their cost." In the next paragraph he warmly exclaims, "Will not the groans, the dying groans, of this deeply afflicted and oppressed people reach heaven? And when the cup of iniquity is full, must not the inevitable consequence be the pouring forth of the judgments of God upon their oppressors?" And in the next, he returns to demographics: "It is a dreadful consideration, as a late author remarks, that out of the stock of eighty thousand Negroes in Barbados, there die every year five thousand more than are born in that island; which failure is probably in the same proportion in the other islands." Finally, he turns to scripture: "Let the opulent planter or merchant prove that his Negro slave is not his brother, or that he is not his neighbor, in the scripture sense of these appellations," he argues, "and if he is not able to do so, how will he justify the buying and selling of his brethren as if they were of no more consideration than his cattle?" Benezet's "scripture sense" derives principally from Mark 3:35, where Christ declares that "whosoever shall do the will of God, the same is my brother, and my sister, and mother," and from Luke 10:25–29, the "Parable of the Good Samaritan," in which Christ clarifies that all humanity are neighbors regardless of faith or ethnicity. The biblical quotations end the chapter, grounding its conclusion in the authority of scripture in a way that structurally balances its opening. In this central chapter of the book, sandwiched between scientific and scriptural authority, Benezet is able to offer his own emotionally freighted response to the slave trade in a place that is both literally and figuratively the heart of the book.[14]

Benezet returns to the Caribbean in the final three chapters of *Some Historical Account.* Once again without Churchill's or Astley's collections to draw from, he blends his own composition with a range of primary and secondary sources, most substantially the anonymous antislavery pamphlet *Two-Dialogues on the Man-Trade,* which had been published in London in 1760 and which itself cited geographies, travelers' accounts, and Sloane's *Voyage to Jamaica.* Benezet's final source is, however, an older one. The argument of the short last chapter is that the Caribbean climate is no impediment to European labor, as some had argued. His evidence comes from Richard Ligon, who was, Benezet notes, a member of the sizable and hard-working white population of the island in the years before enslaved Africans were imported in large numbers and who had noted (in Benezet's paraphrase) that "though the weather was very hot, yet not so scalding but that servants, both christians and slaves, laboured ten hours a day." There are several reasons for Benezet to use Ligon's *History of Barbados* to debunk environmental justifications for slavery based on emerging theories about racial adaptation to climate. Ligon, although sympathetic to the situation of enslaved people in Barbados, is nonetheless clearly no antislavery propagandist and cannot therefore be accused of bias. Furthermore, his observations on climate and population carry weight because they appear as part of an extended natural history of the island, albeit one that was showing its age. This was, however, a further strength. Although more than a century old, Ligon remained an important source for those with an interest in the island. His *History* had stood the test of time and, more importantly, depicted a reality that existed long before slaveholding colonists had consolidated theories about race and climate to justify slavery. Although an account of the Caribbean rather than Africa, Ligon's *History of Barbados* is the final source for Benezet's *Some Historical Account* because it spoke to the underlying and enduring realities of the natural world rather than to the political debates of Benezet's immediate location.[15]

Some Historical Account could not have been written by a Quaker from Pennsylvania who did not travel without extensive reading, and there is no doubt about the texts to which he turned. More than any other genre, natural history writing provided the information that Benezet needed to build a solid case against slavery and the slave trade. The texts he used were in some cases deliberate works of scientific enquiry, such as Hans Sloane's *Voyage to Jamaica* or Griffith Hughes's *Natural History of Barbados.* More often, they were extracts and abridgements of writings by slave traders with an interest in the natural world, particularly those collected and arranged by Awnsham and John Churchill and, especially, by John Green in *Astley's Collection.* Depictions of natural environments, alongside the ethnographic sections of natural histories, were crucial to the antislavery message he formulated and this, in turn, profoundly influenced the emerging genre of abolitionist literature. Indeed, asserts Irv A. Brendlinger, "The most amazing fact about Benezet's writing is the influence it exerted." Benezet assiduously promoted the book and its message and, with the support of what David Brion Davis has called the "Quaker Antislavery International," the text became the first port of call for a new generation of antislavery writers, its tone as well as its content cited, emulated, and imitated in hundreds of pamphlets, letters, sermons, and poems. The "historical account" of its title encompasses both political and natural history, but it is natural history that dominates. In effect, Benezet had produced a natural history of Atlantic slavery that represented the slave trade as a malignant and invasive species, despoiling environments as well as creating untold human misery.[16]

CHAPTER 6

"An unnatural state of oppression"

ENVIRONMENTAL WRITING IN THE ABOLITIONIST ESSAY

After its publication in 1771, Anthony Benezet's *Some Historical Account of Guinea* rapidly became the benchmark for British abolitionist writing, alongside Granville Sharp's 1769 *A Representation of the Injustice and Dangerous Tendency of Tolerating Slavery or of Admitting the Least Claim of Private Property in the Persons of Men, in England.* The works differed in scope, with Sharp's critique of slavery based in a thorough examination of English law and Benezet's drawn from the archive of travelers' and naturalists' accounts of slavery and the slave trade across the Atlantic world. While Sharp's legal arguments were important and frequently cited, particularly in the context of the Somerset case of 1771–72, Benezet's model was more broadly applicable and thus could be—and was—more widely emulated. The impact on abolitionist writing of the 1780s was profound. Numerous authors agreed with Benezet that, while Africa offered resources of great plentitude, the slave trade and the plantation system it served were perversions of the natural order. Three key abolitionist authors built on Benezet's approach, both by making direct use of his work and by reaching more deeply into the same archive of environmental writing. James Ramsay, Thomas Clarkson, and Olaudah Equiano each contributed a milestone text to the abolitionist cause. Ramsay wrote in the mode of an agricultural manual writer. Clarkson made antislavery an academic pursuit. Equiano delivered the eighteenth century's most extensive and compelling slave narrative. These three authors all drew on the work of their immediate predecessors, including one another, as well as the deep archive discussed in the first four chapters of this book. While all also drew on legal, political, philosophical, and scriptural sources, as well as personal experience and private research, the evidence of naturalists and agriculturalists is both explicitly referenced and implicitly invoked throughout their works. Together, they built a compelling case that slavery was unnatural, a case that would in turn inspire tens of thousands of abolitionist activists and authors in the 1780s and beyond.[1]

REWRITING THE MANUALS: THE TREATMENT AND CONVERSION OF SLAVES IN THE BRITISH SUGAR COLONIES

A decade after Benezet stacked up evidence against the slave trade, James Ramsay (1733–89) did the same for the plantations. Described by Folarin Shyllon as "the unknown abolitionist," Ramsay has been overshadowed by others such as Clarkson and Wilberforce largely because he died early in the abolition campaign. For some years in the 1780s, however, he was the most prominent antislavery writer in the public eye, inspiring numerous imitators and some virulent detractors. A Scot trained in medicine at the University of Aberdeen, Ramsay went to sea as a naval surgeon. In 1759, he was called aboard a slave ship afflicted by an epidemic. He was so distraught by the squalid conditions in which the slaves were transported that on leaving the ship he fell and was badly injured. The injury ended his naval career, but the appalling scenes he witnessed set him against the slave trade for life. For almost twenty years thereafter he served as an Anglican priest in St. Kitts, but his uneasiness about slavery and his desire to bring Christianity to his Black parishioners set him at variance with his white parishioners. After years of conflict, he returned to Britain in 1781 and three years later published *An Essay on the Treatment and Conversion of African Slaves in the British Sugar Colonies,* the first of eight books on slavery he issued before his death in 1789. The West India Interest leveled its heaviest guns against him; unlike Benezet and Sharp, Ramsay had been a colonial vicar for two decades, and the group and the planters they represented considered him a traitor. Many public attacks were made against him, of which the most fluent and the most sustained were from the Bristolian James Tobin (1737–1817), formally of Nevis, whom Equiano described as "a zealous labourer in the vineyard of slavery." The violence of the exchange began to tire the reviewers, and may have contributed to Ramsay's early death in 1789, but in the opinion of Thomas Clarkson, it had value in that it "brought Mr. Ramsay into the first controversy ever entered into on this subject, during which, as is the case in most controversies, the cause of truth was spread." Of Ramsay's eight antislavery publications, his first was undoubtedly the most influential. When *An Essay on the Treatment and Conversion of African Slaves* appeared in 1784, Great Britain was still in political turmoil following its defeat in North America. The topic of colonial management—and mismanagement—was therefore firmly in the public eye. Antislavery sentiment was rising too, and while this was in part due to activists such as Benezet and Sharp publishing and campaigning over many years, the British public had also recently been shocked by news of the massacre on the slave ship *Zong* in 1783 in which the ship's captain had murdered 142 captive Africans on board and then attempted to claim back their value as slaves from his insurers. Ramsay's *Essay* arrived, therefore, at precisely the right moment to ignite what until then had been a merely smoldering debate about colonial slavery.[2]

Aligning Ramsay's *Essay* with a specific literary or intellectual tradition is challenging since it is wide ranging and because Ramsay is frustratingly reticent about naming his sources. Much of the book considers the political, theological, and economic history of slavery, and much engages with then current and emerging theories about race. Ramsay is clearly well read in all his subjects, but he adopts a magisterial and encyclopedic tone when discussing intellectual approaches to his topic and only rarely mentions other authors by name. Some parts of the book are grounded in Ramsay's personal practical experience of St. Kitts, but he frequently augments his own knowledge by reference to unspecified acquaintances and correspondents, many of whom are simply referred to decorously, but unhelpfully, as "a gentleman." There are exceptions. His well-known attack in chapter 4 on the racist ideas of David Hume (1711–76) and Henry Home, Lord Kames (1696–1782) explicitly names these fellow Scots, but much of the same chapter is dedicated to a detailed refutation of Edward Long's racial ideas in his 1774 *History of Jamaica,* to whom Ramsay politely refers only as "the ingenious author of a late History of Jamaica." While Ramsay's reading can sometimes be easily inferred, as here, more often his encyclopedic tone combined with his parsimonious referencing make it difficult to identify his underlying sources. The book's central argument is, however, that slavery is unnatural, while several chapters respond to and even parody plantation management manuals of the sort produced by Samuel Martin. Despite Ramsay's attempts to efface his own influences, therefore, large parts of the *Essay* can be located within the tradition of environmental writing about the plantation.[3]

We do have some clues to Ramsay's reading. As Shyllon has shown, Ramsay explicitly challenged Long's theory that African people and orangutans were closely related in a letter written to the *European Magazine* in 1788, although on this occasion Ramsay wrote anonymously. In his 1785 reply to the proslavery response to his *Essay*, Ramsay refutes attempts to appropriate Richard Ligon's *History of Barbados* as evidence for the planters' claim that sugar could not be grown without imported slave labor. In 1786, Ramsay edited and published a letter written by John Samuel Smith (fl. 1786) to which he added a long introduction and postscript. Evidently wounded by criticism that he had not named his sources, Ramsay defended his practice: "Did I think it necessary, I might corroborate the abundant proofs given in the Essay, by the testimonies of many living witnesses, and by quotations from a multitude of authors of undoubted credit. . . . But I really did not conceive it to be incumbent on me to adduce such an accumulation of evidence in support of a fact too plain to be controverted. It seemed to me the same thing as going about to prove, by a long train of formal arguments, that *it was light at noon day*." Nevertheless, he adds a substantial footnote naming eight key texts, including Sloane's *Voyage to Jamaica* and Benezet's *Some Historical Account of Guinea*.[4]

In addition, surviving manuscript notes in the Bodleian Library and the British Library show that he read widely and kept extensive notes from a large number of authors concerned with slavery as well as those interested in Africa and the Caribbean more generally. Many of these were naturalists or those interested in natural history. For example, Ramsay took two densely written folio pages of notes from "Ligon Hist of Barbados 1647 [*sic*]." Ramsay pays particular attention to Ligon's account of the environmental transformation of the island. In Ramsay's synopsis, Ligon reports, "When Island discovered men and tools were sent from England to cut down woods & clear the ground and once beginning to taste the sweet of the trade they set hard to work and lived in much better condition than at first." Ramsay also takes almost verbatim notes of the stories of Macaw and the theorbo, Sambo who wished to be baptized, and Ligon's story of an "Indian maid," to which he adds the observation "Woman is the Yarico of the Spectator." In the fifth section of chapter 4 of the *Essay*, Ramsay had presented three anecdotes to illustrate "African capacity" although, he remarks, "a volume might easily be filled." These manuscript notes are clearly part of that unpublished volume. In addition to those taken from Ligon, Ramsay also took extensive notes from other natural history texts, including Long's *History of Jamaica*, "Nicholsons Nat. Hist of St Domingo Printed at Paris 1776," and "Extracts from Mr Henry Smeathman's manuscript on Slavery and the African trade Letter." The surviving manuscripts appear only to be a fraction of all the notes that Ramsay took while preparing his *Essay* and other publications, but they, along with the footnote to Smith's letter, clearly demonstrate that his thinking was profoundly influenced by a long tradition of environmental writing.[5]

The *Essay on the Treatment and Conversion of African Slaves* begins by arguing that slavery is unnatural. The opening line acknowledges that "there is a natural inequality, or diversity, which prevails among men that fits them for society" but maintains that in a functioning society "the feelings and interests of the weaker, or inferior members, are consulted equally with those of the stronger or superior." While "each man takes that station for which nature intended him," his rights and claims "are restrained, by laws prescribed by the Author of nature." This rosy conception of a harmonious society might have surprised the poorest of his parishioners, who were no doubt rarely consulted equally, but it was a conventional enough view, if somewhat complacent, for an Anglican clergyman. "Opposed to this law of nature," argues Ramsay, "stands the artificial, or unnatural relation of master and slave." He explores this over the ensuing fifteen pages before concluding that slavery "is an unnatural state of oppression on the one side, and of suffering on the other; and needs only to be laid open or exposed in its native colours, to command the abhorrence and opposition of every man of feeling and sentiment." The book that follows is just such an exposé, written in the expectation that readers will naturally turn against what is unnatural just as soon as they know the facts. Although this is an argument based in natural law rather than natural history, it nonetheless accords with representations of slavery that saw it as inconsistent both with God's plenitude in nature and the notion that he "hath made of one blood all nations of men" (Acts 17:26). All later arguments in the book must, therefore, be understood within this conceptual framework.[6]

One way to understand the book is as an alternative plantation management manual. Ramsay claims in his preface that the *Essay* had begun as a personal letter but "by frequent transcription, it sensibly increased in size, and extended itself to collateral subjects, till it had become something like a system for the regulation and improvement of our sugar colonies." His thoughts on the plantation environment mostly come in a section of the first chapter called "Master and Slave in the British Colonies," where he describes current plantation conditions, and in the last section of the final chapter, where he offers a scheme for future amelioration. Ramsay's "system" is thus also a framing device, within which he can offer a broader intellectual history of slavery while always reminding the reader that the issue starts and ends with the reality of enslaved people laboring on a plantation. His system is, nevertheless, a practical guide to managing a reformed plantation that presents the subject matter and follows the formal requirements of the genre. Like Ligon, Martin, and others who had written on plantation management, Ramsay observes that the principal business of the plantation is "to manure, dig, and hoe, plow the ground, to plant, weed, and cut the cane, to bring it to the mill, to have the juice expressed, and boiled into sugar." This is the daily grind of plantation life, and the central business of a plantation manual, but Ramsay's purpose is to expose as well as to regulate and improve. Where Martin had proudly portrayed what he believed to be a modern, rational, and humane system, Ramsay instead exposes the horrors that arise, not only from the lash, but also from the planters' approach to the environment itself. The worst offender is grass. Martin had stressed that "cattle require change of food to preserve them in health" and that one foodstuff was "a variety of grass, which every soil produces with a little care in moist weather." Ramsay expands at length on what "a little care" really means by explaining that "the slaves are dispersed in the neighbourhood, to pick up about the fences, in the mountains, and fallow or waste grounds, natural grass and weeds for the horses and cattle. The time allotted for this branch of work, and preparation of dinner, varies from an hour and a half, to near three hours. In collecting pile by pile their little bundles of grass, the slaves of low land plantations, frequently burnt up by the sun, must wander in their neighbours grounds, perhaps more than two miles from home." Failure to collect enough grass results in twenty lashes, and some, says Ramsay, rob others of their bundles to avoid the labor. "This picking of grass," he says, "is the greatest hardship that a slave endures" but is quite unnecessary; "a few acres of land, in proportion to the extent of the plantation, allotted for artificial grass, and a few weakly slaves separated from the work, would take away the necessity of providing for cattle in this harrassing scanty manner."[7]

Although Ramsay comprehensively demolishes Martin's notion of "a little care," and it is tempting therefore to see the passage as a direct response to Martin, Ramsay's habitual effacement of his sources, and his refusal to embarrass any planter by mentioning them by name, means that we cannot be certain whether Ramsay has Martin in mind. It is highly likely, however. Ramsay would almost certainly have read Martin's *Essay upon Plantership*, given that it was the most substantial guide to running a plantation available while Ramsay was himself a planter. The two men may even have been personally acquainted, for better or for worse. St. Kitts is a mere eighty kilometers (fifty miles) from

Antigua, and in a manuscript letter to an unknown correspondent about his edition of Smith's *Letter from Capt. J. S. Smith,* Ramsay mentions that his knowledge of the Caribbean "extended in fact from Barbados down to the Virgin Islands" and that he thought of Antigua as "a colony which prides itself on its superior degree of polish and police." This may be an oblique reference to Martin. He was speaker of the Antigua Assembly and colonel of the island's militia at the same time that Ramsay was a vicar in St. Kitts and was known both for his polished hospitality to visitors and his pride in showing them his plantation, as we saw earlier in the account by Janet Schaw, whom Martin entertained in December 1774. If Martin ever entertained Ramsay, it is not immediately obvious from the approximately three hundred thousand sprawling words of his surviving handwritten letters. Martin did write a testy letter in May 1766 to a certain "Mr. Ramsey" who had threatened to sue Martin over an unpaid debt from a "Mr. Stewart" for which Martin had acted as guarantor, but this Ramsey appears to be an Antiguan farrier rather than the St. Kitts' clergyman. Nevertheless, it is inconceivable that Ramsay was not aware of the famous Martin and his work. We can therefore have confidence in reading the opening and closing chapters of Ramsay's *Essay* as a response to Martin's *Essay upon Plantership.*[8]

Ramsay frequently tackles the same topics as Martin but, unlike Martin, is at pains to show the impact plantation activities have on enslaved laborers. For instance, Martin had spoken at length about the importance of dunging, advising, "When any of these soils are exhausted of their fertility, by long and injudicious culture, they may be restored by any kind of dung well rotted." Ramsay appears to challenge the nonchalance of Martin's prescription, revealing how this dung is obtained and how far from easy or natural the rotting process is. The compost is formed from that very grass collected with great labor by enslaved people in their supposed free time:

> This grass, except such part of it as is reserved for the stable horses, procured by so much toil, and forced out of the slave by such repeated punishment, under pretence of feeding the cattle and mules, is spread abroad under their feet, on a fermenting inclosed dung heap, called a pen. There a very considerable part is lost to every purpose of nourishment, by being trampled under the beasts feet; where mixing with dung and urine, it ferments, corrupts, and with its suffocating steams in that sultry climate, instead of supplying them with vigour, fills them with disease; as if Providence meant to revenge the oppression of

> the slave, in being forced to drudge thus for it, by inspiring the master with a spirit of absurdity, in his manner of using it.

Ramsay emphasizes the cruelty and hypocrisy of the planters, who use "repeated punishment" to enforce work that they only pretend is directly useful, a process he then undercuts by layers of irony. Most of the nourishment of the grass is lost and, instead of feeding the animals, it "fills them with disease." The passage ends with an act of divine revenge that bathetically renders the planters absurd. The activities of grass picking and manure production that Martin claimed could be managed with "a little care" are exposed as cruel and ineffective. The passage is an unambiguous satire of both the technique and the literature of plantation management.[9]

A similar method can be observed in Ramsay's discussions of marginal planting and vegetable gardens, both of which are similar to Martin's in content and argument but very different in tone. Both authors decry absentee landlords and the encroachment of sugar monoculture into margins and gardens as well as the dangers of continual cropping without fallowing or rotation. In his section on husbandry, for example, Martin emphasizes the importance of planting roadside trees for shade, both for cattle and people, since "if the care of providing shade for brute creatures is so much the duty and interest of their owners, how much more agreeable to the laws of humanity to provide shade for every human creature travelling upon the high roads in this hot climate?" He notes the frequent economic argument that the roots of the trees will cause "injury to canes" but dismisses this as "too mean to deserve an answer: but to gratifie the muckworm, let him dig a small trench between his canes and trees." Martin and Ramsay would probably have agreed on this point, but Martin does not go on to discuss the benefits of marginal planting to the enslaved people on the plantation. Instead, he launches into a rhapsody on the potential beauty of the island which, if it "were planted with avenues along all the high roads; and the summit of every barren hill crowned with clumps of trees, it would be the most magnificent garden the world could ever boast of since that of Eden." This image of a tropical paradise elides the fact that, for its enslaved laborers, the island was probably as close to hell on Earth as could be imagined. Martin concludes, characteristically, by noting the economic benefits to the planters: "That very beauty might not only render it more healthful to the inhabitants, by preserving them from fevers kindled by the burning sun-beams; but also much more fruitful, by seasonable weather: for, as by cutting down all its woods, an hot country becomes more subject to excessive droughts, so by replanting it in the manner above described, it would probably become more fruitful." Martin acknowledges the effect on local climate of deforestation, a problem that was apparent from the mid-seventeenth century onward, but his solutions, as ever, are for the benefit of the planter.[10]

Ramsay does not discuss the shade benefits of trees directly, although he does note that "the slaves of low land plantations, frequently burnt up by the sun, must wander in their neighbours grounds, perhaps more than two miles from home" in search of bundles of grass. His discussion of the advantages of marginal planting and crop rotation instead stresses the importance of such plots for growing food. This had concerned Martin as well, and the two passages have much in common. Martin argues that some of a planter's "most fruitful land should be allotted to each negro in proportion to his family, and a sufficient portion of time allowed for the cultivation of it." This, he argues, engenders both physical and economic health since "if the labor of producing our own provisions was fairly computed, and compar'd to the expence of purchasing that of North America, I dare affirm the latter will be found as expensive, tho' much less wholesome and nutritious." Neglecting these allotments is a cause of Antigua's "general poverty," and he concludes, with a nod to effective crop rotation, that "he therefore who will reap plentifully, must plant great abundance of provisions as well as sugar canes: and it is nature's œconomy so to fructify the soil by the growth of yams, plantains, and potatoes, as to yield better harvests of sugar." Both authors decry the false economy that plants up field margins, fallow land, and allotments with sugarcane, but where Martin offers open encouragement to his fellow planters to diversify their crops, Ramsay presents his arguments satirically. "Formerly," he notes, "before we became such accurate planters, and before luxury had rapaciously converted every little nook of land into sugar, the slaves had a field or two of the fallow cane-land yearly divided among them, for a crop of yams, pease, and potatoes." Where Martin derides the individual "muckworm," or miser, Ramsay sees a deeper, more structural problem: "The practice of turning all our lands to the growth of the sugar cane, and neglecting the culture of provisions for the slaves, and of artificial grass for the cattle, has lately arisen equally from the demands of extravagance in our absent planters, and of poverty in those on the spot. Sugar, sugar, is the incessant cry of luxury, and of debt. To increase the quantity of this commodity, gardens of half an acre have been grubbed up; and that little patch, which he had used to till for his own pease, or cassava, has the slave been made to dig for the reception of his master's sugar cane." Like Martin, Ramsay sees this as a false economy. Unlike Martin, who had merely asserted that it is "the duty of a good planter to inspect every part of his plantation with his own eyes," Ramsay launches into a vitriolic attack on absentee planters who, through ignorance and indifference, increase both the economic and personal costs at the plantation for nothing better than a small uptick on their balance sheets back in England:

> Hence the annual expence of plantations, within less than thirty years, has been more than doubled. Hence the sending of two or three extra casks of sugar to market has been attended with an expence of hundreds of pounds in provisions to slaves, in oats to horses, and in keeping up the stock of slaves and cattle, worn out, before their time, by indiscreet extraordinary efforts, and a scanty allowance. The peculiar fertility of St. Christopher's has the most baneful effects. It enables the greatest part of its proprietors to live in England; where, insensible of the sufferings of their slaves, they think and dream of nothing but sugar, sugar; to which, in consequence, every spot of land is condemned.

Both Martin and Ramsay perceive the same financial and agricultural issues and, at the plantation level, both advocate similar solutions of rotation, diversity, and self-sufficiency to promote sustainable cropping and improved long-term yields, even if at the expense of some short-term profits. Where they differ is in their attitude to the enslaved people whose labor powers the plantation economy. Martin is a classic ameliorationist, accepting the system as a whole while encouraging individuals to adopt local and partial reforms. Ramsay also advocates reform, but he does so through satirical passages such as these that expose the absurdity of ameliorationist positions even as they advance them, and that call into question the underlying economic model on which the plantations are premised.[11]

Ramsay's immediate aim in the *Essay* is ameliorative rather than emancipatory. Although he would almost immediately afterward take a more robust stand against the slave trade, the *Essay* itself calls only for reform and his vision is in many passages remarkably congruent with Martin's conception of a well-ordered plantation. Ramsay's idea of an improved plantation would be unlikely to impress many modern environmentalists or humanitarians. His solution to the grass problem, for example, calls for "a few acres of land, in proportion to the extent of the plantation, allotted for artificial grass." In other words, he is calling for an intensification of agriculture on an already overcultivated island, sowing a grass monoculture of an introduced species where previously "natural weeds" had flourished. The idea of giving the labor of this fodder pasture over to "a few weakly slaves" is hardly a radical blow for freedom. Indeed, it should not be supposed that Ramsay's plan for amelioration, as it appears in the final chapter of the *Essay*, is in any sense a call for emancipation, still less for self-determination. Ramsay's reformed planter remains a manager of the estate, both its lands and its enslaved people, exercising "soft power" of a sort that not even Martin had proposed. Ramsay's management plan calls for minute control over the daily life of captive Africans and the replacement of their cultures and beliefs with European modes. The agent of much of this control is the Anglican Church. Ministers should educate enslaved children so that "while their minds are tender, before their dispositions be soured by the impositions of slavery, they may make some progress in the knowledge of their duty." As adults, "Negroes, who are well treated and in spirits, sing at work. A few easy single stanzas might be collected or composed, to be used instead of their common songs." Ministers should "visit the plantations in rotation, at convenient times, to inquire into the behaviour and improvement of the slaves, to commend, reprove, admonish, and pray with them. To give him respect and influence, let all be obliged to appear before him decently clothed." These are rules of the sort by which Anglican missionaries asserted British cultural power across the British Empire and, indeed, in 1787 Ramsay himself issued a small forty-page pamphlet called *A Manual for African Slaves* that offered catechism and "easy stanzas" intended to evangelize enslaved Africans.[12]

While Ramsay certainly advanced the idea that Caribbean plantations were unnatural, mismanaged, cruel, and therefore in need of reform, the success of his *Essay* paradoxically familiarized British readers with the detail of plantation life in its exotic tropical setting, reinforcing rather than challenging the notion that the West Indies could, and should, be subject to British colonial management. But for all that, it was nonetheless a wake-up call, alerting many in the British reading public for the first time to the inhumanity of plantation slavery. Its effectiveness derived in part from Ramsay's own personal authority as one who had been within the plantation system, but in more literary terms its blend of satire and sensibility and its adoption in key sections of the form of a plantation management manual lent it both the authority and the authenticity it needed to open a widespread public debate that began almost immediately when Ramsay was embroiled in a pamphlet war, first with opponents from his former parish and then with the full weight of the West India Interest in Britain. From this point forward, the problem of slavery would occupy a central place in British political discourse and it was to Benezet's and Ramsay's depictions of slavery as unnatural, both built on the archive of environmental writing in the service of the plantocracy, to which campaigners would first turn.

THOMAS CLARKSON, ON THE SLAVERY AND COMMERCE OF THE HUMAN SPECIES

In 1785, at the height of the pamphlet war arising from James Ramsay's exposé of plantation slavery, students at Cambridge University were asked to write a prize essay on the question "Is it lawful to make slaves of others against their will?" Thomas Clarkson was eager to win the prize and his research led him quickly to the writings of Anthony Benezet. Winning the competition convinced Clarkson to make antislavery his life's work and he published a revised and augmented version of the essay in London in 1786 after having met with William Dillwyn (1743–1824), a Pennsylvanian former student of Benezet now spreading the antislavery message in London, and James Ramsay, who traveled up to London from Kent specially to meet with him. Clarkson's *Essay on the Slavery and Commerce of the Human Species, Particularly the African* met with immediate public interest and, augmented by his efforts out on the campaign trail, was vital to the success of the Society for Effecting the Abolition of the Slave Trade in the 1780s. Clarkson would continue to campaign against slavery into the 1830s, albeit with periods of burnout brought on by his punishing schedule. Recognized today as the powerhouse of the late eighteenth-century abolition movement, Clarkson was also its first historian, publishing in 1808 a monumental *History of the Rise, Progress, and Accomplishment of the Abolition of the African Slave Trade.* Although Clarkson's *Essay* quickly became a centerpiece of antislavery literature and is regularly mentioned by historians of abolitionism, it has received surprisingly little critical attention, perhaps because of what Adam Hochschild has called "a certain mustiness of style," albeit balanced by its "heartfelt outrage." The mustiness perhaps derives from its origins as a student essay written in Latin, but this does not appear to have affected its popularity: it was reprinted several times and a second edition of 1788 considerably revised and expanded the first. The book's central philosophical point is that slavery is unnatural because the natural condition of humanity is freedom. "It appears first," Clarkson plainly states, "that *liberty* is a *natural,* and *government* an *adventitious* right, because all men were originally free," and this argument is repeated in various forms throughout the book.[13]

Clarkson's view is informed by deep reading, and while many of his sources are philosophical, legal, or historical, he also mined the works of natural historians and engaged with contemporary biological theory. Much of his approach derives, however, from the extensive use he made of Benezet's *Some Historical Account of Guinea* when preparing the original Cambridge University essay. The book begins with fulsome praise for the Quaker position on slavery, singling out Benezet's work in particular, but Benezet's influence manifests itself in less direct ways throughout the *Essay.* For example, drawing on and rhetorically augmenting Benezet's discussion of African river systems, Clarkson argues that "the unfortunate Africans, terrified at these repeated depredations, fled in confusion from the coast, and sought, in the interiour parts of the country, a retreat from the persecution of their invaders. But, alas, they were miserably disappointed! There are few retreats, that can escape the penetrating eye of avarice. The Europeans still pursued them; they entered their rivers; sailed up into the heart of the country; surprized the unfortunate Africans again; and carried them into slavery." Clarkson concludes the passage by emphasizing both the environmental and the human devastation this strategy causes as well as showing that it is not sustainable. "The banks of the rivers were accordingly deserted, as the coasts had been before; and thus were the *Christian* invaders left without a prospect of their prey." The metaphorical as well as the practical importance of rivers would remain an important theme throughout Clarkson's literary career, perhaps most famously in his 1808 visual representation of the abolitionist movement in which individual abolitionists are shown as a series of tributaries flowing into the rivers and estuaries of abolitionism (see figure 1). The inspiration for this idea appears to have been Benezet, and throughout the *Essay* we can identify other more or less direct points of reference to Benezet's work. In his discussion of the fecundity of the African soil, for instance, Clarkson imitates Benezet (who was himself quoting from Michel Adanson) when he notes that "Africa is infested with locusts, and insects of various kinds; that they settle in swarms upon the trees, destroy the verdure, consume the fruit, and deprive the inhabitants of their food. But . . . the very trees that have been infested, and stripped of their bloom and verdure, so surprizingly quick is vegetation, appear in a few days, as if an insect had been utterly unknown." In these passages, Clarkson incorporates Benezet's interest in the effect of slave trading on the African environment almost verbatim.[14]

Benezet was Clarkson's primary source for the student essay, and his influence on the published *Essay* is evident throughout, but by the time the published work appeared, Clarkson had expanded his reading considerably to include, as he enumerates in his preface, "Sir *Hans Sloane's* Voyage to Barbadoes; *Griffith Hughes's* History of the same island, printed 1750; an Account of North America, by *Thomas Jefferies,* 1761; all *Benezet's* works, &c. &c. and particularly . . . Mr. *Ramsay's* Essay on the Treatment and Conversion of the African Slaves in the British Sugar Colonies." Clarkson is far more explicit about naming his sources than Ramsay, but he alternates between adopting Benezet's method of referring directly to multiple sources, often in long quotations, and Ramsay's technique of synthesizing his sources into a seamless narrative in which the originals are decorously obscured. Clarkson's history of African slavery in part 1, chapter 8 is a good example of the former approach. The section is introduced with a footnote explaining, "The following short history of the African servitude, is taken from Astley's Collection of Voyages, and from the united testimonies of Smyth, Adanson, Bosman, Moore, and others, who were agents to the different factories established there; who resided many years in the country; and published their respective histories at their return. These writers, if they are partial at all, may be considered as favourable rather to their own countrymen, than the unfortunate Africans." Whether Clarkson came to *Astley's Collection* independently or was introduced to it by Benezet is unclear, but the method is undoubtedly Benezet's and the passages that follow replicate, as we have seen, Benezet's interest in river systems, an interest that itself reflects the physical realities of European trade and exploration in West Africa. Clarkson does not, however, merely confine himself to Benezet and his sources. The same chapter contains testimony from naturalists whose work was not available to Benezet. For instance, Clarkson includes a long passage from the "learned author" Anders Sparrman (1748–1820), the celebrated pupil of Linnaeus and "professor of Physick at Stockholm, fellow of the Royal Academy of Sciences in Sweden, and inspector of its cabinet of natural history." In this extract from his account of his scientific expedition to southern Africa in the 1770s, Sparrman describes "the method which the Dutch colonists at the Cape make use of to take the Hottentots and enslave them." The slaughter is horrific. If a colonist sees a Khoikhoi person at any time, says Sparrman, "he takes fire immediately, and spirits up his horse and dogs, in order to hunt him with more ardour and fury than he would a wolf, or any other wild beast." Clarkson ends the chapter by concluding that this behavior is unnatural rather than merely bestial. "The lion does not imbrue his claws in blood," he argues, "unless called upon by hunger, or provoked by interruption; whereas the merciless Dutch, more savage than the brutes themselves, not only murder their fellow-creatures without any provocation or necessity, but even make a diversion of their sufferings, and enjoy their pain." Hunting for food may be natural, but sadism is not.[15]

In part 3 of the *Essay,* Clarkson's method more closely resembles Ramsay's in that it conflates and elides its sources, but Clarkson takes the technique to another level by choosing to "throw a considerable part of our information on this head into the form of a narrative," which, "though it may be said to be imaginary, is strictly consistent with fact." He then follows the journey of an imaginary captive African as he is taken to the coast, put aboard a ship in the Middle Passage, and sold in the Caribbean. As Vincent Carretta has shown, Clarkson was criticized by the proslavery apologist Gilbert Francklyn (1733–99) for "inventing black witnesses against slavery—witnesses, Francklyn insinuates, who were actually white men speaking in blackface." Francklyn challenged Clarkson "to produce a Negro of character who would not *turn pale* in fabricating such assertions. I call upon Mr. Clarkson to produce any book he ever perused, even Mr. Ramsay's (whose accounts are completely proved to be untrue by Mr. Tobin), in which he found such stories related." Carretta notes that Olaudah Equiano would shortly provide exactly the account from "a Negro of character" that Francklyn demanded, but in any case, Clarkson's narrative did in fact rest on several books that he could easily have produced, including, again, "Astley's Collection of Voyages, and . . . the united testimonies of Smyth, Adanson, Bosman, Moore, and others," refracted at least in part through Benezet. His account of life on a Caribbean plantation is clearly influenced by Ramsay because of its emphasis on details such as "the most laborious and intolerable employment" of "the collection of grass for cattle." But perhaps the most important and lasting influence on Clarkson's thinking was the emphasis placed by natural historians of Africa on the richness and plentitude of the African environment. "Nothing can be more clearly shewn," Clarkson argued in the *Essay,* "than that an inexhaustible mine of wealth is neglected in Africa, for the prosecution of this impious traffick; that, if proper measures were taken, the revenue of this country might be greatly improved, its naval strength increased, its colonies in a more flourishing situation, the planters richer, and a trade, which is now a scene of blood and desolation, converted into one, which might be prosecuted with advantage and honour." This belief animated Clarkson throughout his career, such that from the following year onward, he began to travel with his famous case of African natural produce to demonstrate, in his public speaking engagements as well as to Parliament, the riches that the continent might yield.[16]

If Clarkson was clearly influenced by those who had described African and Caribbean environments, he also engaged with the ideas of those who were attempting to theorize racial difference. His use of the term "human species" in the title plainly signals that he rejects any suggestion that Africans and Europeans are separate species. He expands on this at length by debunking what he calls that "system of reasoning" in which it was argued that "the Africans are an inferiour link of the chain of nature, and are made for slavery." Some observers, such as the notorious Edward Long, saw a hierarchical gradation from Europeans to Africans to orangutans. Clarkson explicitly rejects both this assertion and the biological theory that underpins it:

> It is an universal law, observable throughout the whole creation, *that if two animals of a different species propagate, their offspring is unable to continue its own species.* By this admirable law, the different species are preserved distinct; every possibility of confusion is prevented, and the world is forbidden to be over-run by a race of monsters. Now, if we apply this law to those of the human kind, who are said to be of a distinct species from each other, it immediately fails. The *mulattoe* is as capable of continuing his own species as his father; a clear and irrefragable proof, that the scripture account of the creation is true, and that "God, who hath made the world, hath made of one blood all the nations of men that dwell on all the face of the earth."

The scripture quotation, from Paul's Areopagus Sermon (Acts 17:26), confirms that Clarkson is a monogenist, a believer that only one species of humanity had been created by God. He also strongly asserts the belief in distinct natural species, thereby rejecting the infinite gradations of the great chain of being. "But if this be the case," he continues, "it will be said that mankind were originally of one colour; and it will be asked at the same time, what it is probable that the colour was, and how they came to assume so various an appearance?" His explanation is offered in a chapter where, as Manisha Sinha puts it, "he summarized the climatic theory of race popularized by eighteenth-century environmentalists." Clarkson suggests that, over generations, populations gradually darken or lighten in color as a response to their local climate, and he cites a tendency of Africans in the New World to give birth to lighter-colored children. Hochschild points out that Clarkson was displaying considerable naïveté in not being aware that rape was a widespread manifestation of plantation brutality. Nevertheless, despite his error, this is a proto-evolutionary assertion that populations adapt to their environment over time. Clarkson suggests that the original color must have been that "of Noah and his sons, from whom the rest of the world were descended" which, as they were in the Mediterranean region, was likely to be "a dark olive; a beautiful colour, and a just medium between white and black. That this was the primitive colour, is highly probable from the observations that have been made; and, if admitted, will afford a valuable lesson to the Europeans, to be cautious how they deride those of the opposite complexion, as there is great reason to presume, *that the purest white is as far removed from the primitive colour as the deepest black*." The debate about human origins and human variation was in its infancy in the 1780s. The ideology of what would become known as "race" was for the first time being theorized in the pseudoscientific terms in which it would be asserted in increasingly pernicious and violent terms in the nineteenth and twentieth centuries. In the face of emerging racial ideologies, and using the theories available to him at the time, Clarkson argues that "complexion" was merely a minor and transitory element of a human species that in all other respects was, as Paul had put it, "of one blood." Like other abolitionists, Clarkson blends evidence from scripture and natural history in a way that is consistent with eighteenth-century natural theology, but that also furthered and popularized the emerging ideas of both race theorists and proto-evolutionists. Clarkson's decision to deploy an environmental explanation for human diversity arises from a long debate with classical antecedents that had recently found new expression in the work of Montesquieu (1689–1755), David Hume, and others, but his choice would influence many later abolitionist authors. By the time abolitionists came to write their own history, it seemed only natural to represent the rise, progress, and accomplishment of the abolition of the African slave trade as an environmental feature: a series of tributaries and rivers flowing to the great sea of abolition. The famous map of the abolition movement that folds out from Clarkson's 1808 history of the movement was a fitting testimony to the entangled discourses of antislavery and natural history.[17]

"AN INEXHAUSTIBLE SOURCE OF WEALTH": OLAUDAH EQUIANO AND AFRICAN NATURAL HISTORY

No one of African birth or descent is mentioned on Clarkson's map of "abolitionist forerunners," unless any of the three references to "Anonymous" meant an African abolitionist. Clarkson was nevertheless not ignorant of

African contributions both to English literary culture and to the abolition movement. He reproduces three poems by Phillis Wheatley (c. 1753–84) in his *Essay*, followed by a discussion of the letters of Ignatius Sancho (c. 1729–80), backed up by a statement that the equal capacity of African and Europeans "has been constantly and solemnly asserted by the pious Benezet." We do not know if Clarkson was familiar with *Thoughts and Sentiments on the Evil and Wicked Traffic of the Commerce of the Human Species*, the first explicitly abolitionist publication in English by an African, which was published in 1787 in London by Ottobah Cugoano (c. 1757—c. 1791). We do know, however, that Clarkson was on good terms with Cugoano's friend Olaudah Equiano, since Clarkson wrote Equiano a letter of introduction to the Reverend Thomas Jones (1756–1807), master of Trinity College, Cambridge, asking Jones to welcome "a very honest, ingenious, and industrious African, who wishes to visit Cambridge." The introduction into Cambridge society was fortuitous for Equiano as it led to him meeting his future wife Susanna Cullen (1759–96): a romance in which Clarkson thus played an unexpected role.[18]

Despite uncertainty about Equiano's birthplace, his life story is now well known. By his own account he was born in West Africa around 1745, but external evidence suggests he may instead have been born in South Carolina—a mystery that has given rise to a lively scholarly debate. Whatever the truth about his very early years, it is undisputed that by the mid-1750s he was enslaved and serving in the Royal Navy during the Seven Years' War, after which he was sold to a planter in the Caribbean island of Montserrat from whom he was able to purchase his freedom in 1766. He worked as a seaman for the next twenty years before joining the abolition movement and publishing his autobiography, *The Interesting Narrative of the Life of Olaudah Equiano, or Gustavus Vassa, The African*, in 1789. The last years of his life were spent on the campaign trail until his death in 1797. Popular in its day and into the 1830s, then forgotten for a century and half, *The Interesting Narrative* is today celebrated as an important firsthand account of a journey from slavery to freedom—but also as one of the most compelling autobiographies of its age.[19]

Recent scholarly inquiry into Equiano's life and work has been extensive and varied. Some have investigated *The Interesting Narrative*'s relationship with place. Elizabeth A. Bohls has noted that Equiano's "literary and business acumen led him to package his experience as travel writing. Travel writing—the literature of place—brings geography to life through first-person narrative." Equiano's vivid engagement with many places is, no doubt, one of the many reasons why his *Narrative* continues to be read today. By contrast, relatively few scholars have paid attention to his engagement with the natural world, although Justin Hosbey, Hilda Lloréns, and J. T. Roane consider it notable that Equiano "marks place and movement through references to taste and nature." They argue, "This alternative mode of recalling place and environment suggests the enduring Africanness of Black immersion and movement through and with the environment, one distinct from the dominant visual power associated with Enlightenment and enclosure." Their argument is based largely on the opening chapters of the *The Interesting Narrative,* in which the author describes a childhood in rural West Africa. Much of the rest of the book, however, like Equiano's adult life, is either maritime or urban, while his reading interests tended to be either literary or spiritual rather than scientific, and, reading and writing aside, his leisure hours seemed mostly occupied with music or the church. Nevertheless, he does take a sustained interest in his environment, not only in the accounts of his childhood in the first three chapters, but also in the arguments against slavery at the end of the book. These important but overlooked engagements with the natural world can be divided into two categories: those that Equiano explicitly signals as inspired by Benezet and others and those that he offers as unique personal experiences. Taken together, these passages not only attest to the deep penetration of natural history writing into abolitionist discourse, even in the writing of an author with little apparent interest in the topic, but also offer us additional ways of thinking about the thorny question of Equiano's birth and early life.[20]

The first chapter of *The Interesting Narrative* offers "the author's account of his country," by which he refers to "a charming fruitful vale, named Essaka" in the kingdom of Benin, today part of southern Nigeria. Despite speculation, the location of this valley has never been unambiguously identified. Equiano's description of Essaka is principally ethnographic, but it begins, as do many natural histories, with its geographical location and a brief description of the physical environment, in this case, the "the richness and cultivation of the soil." Equiano returns to this a few pages later in a longer passage that invokes the depictions of African prodigality found in Churchill's and Astley's collections and reworked by Benezet and Clarkson. In Equiano's formulation, "Our land is uncommonly rich and fruitful, and produces all kinds of vegetables in great abundance. We have plenty of Indian corn, and vast quantities of cotton and tobacco. Our pine apples grow without culture; they are about the size of the largest sugar-loaf, and finely flavoured. We have also spices of different kinds, particularly pepper; and a variety of delicious fruits which I have never seen in Europe; together with gums of various kinds, and honey in abundance.

All our industry is exerted to improve those blessings of nature." Later, when the captive child is brought nearer to the coast, he notes that "the soil was exceedingly rich; the pomkins, eadas, plantains, yams, &c. &c. were in great abundance, and of incredible size." Equiano's descriptions of African plentitude match those in Benezet and his sources, as do other details of his journey to the coast, such as his depiction of the use of river systems or his account of a plague of locusts. The emphasis on prodigality also aligns Equiano with the argument made by Benezet and Clarkson that Africa was a "mine" of valuable natural resources that might be exploited were the slave trade not inhibiting its development. This becomes explicit in the closing pages of the book where Equiano notes that Africa is "a continent, nearly twice as large as Europe, and rich in vegetable and mineral productions" and that "the bowels and surface of Africa, abound in valuable and useful returns." The debt to Benezet and Clarkson is clear, but it is not in dispute since Equiano explicitly cites both authors in his footnotes. The pertinent question is why Equiano would want to call upon these authors rather than speaking entirely from personal experience. Critical responses tend to reflect critics' positions on the birthplace question, siding either with Carretta's view that Equiano "may have invented rather than reclaimed an African identity" or Paul E. Lovejoy's observation that Equiano "never claimed that the details of the Bight of Biafra were entirely based on his own experiences" but were imperfect childhood memories augmented by discussions he had with other "natives of Eboe" and "quotations from Benezet and other sources." Whether or not Equiano constructed or reconstructed an African childhood, however, the rhetorical maneuver remains the same. He is framing his narrative with a depiction of Africa and an argument about African natural productivity that was first marshaled by Benezet from the works of naturalists and later deployed by Clarkson in his *Essay.* By combining Benezet's encyclopedic method with Clarkson's narrative technique, Equiano signals that his *Interesting Narrative* is aligned with the key texts of abolitionist discourse, a strategic choice that could only have strengthened its authority in the debate.[21]

Fewer critics have paid attention to Equiano's personal engagements with the natural world, perhaps since they appear to speak less to the contention over his birthplace or because they are not obviously related to his testimony against the slave trade. Many are accounts of creatures encountered on or near the ocean, and there is relatively little about African fauna in his depiction of Essaka. A notable but rather curious exception is his description of an incident that took place when he would have been about seven or eight years of age.

> We have serpents of different kinds, some of which are esteemed ominous when they appear in our houses, and these we never molest. I remember two of those ominous snakes, each of which was as thick as the calf of a man's leg, and in colour resembling a dolphin in the water, crept at different times into my mother's night-house, where I always lay with her, and coiled themselves into folds, and each time they crowed like a cock. I was desired by some of our wise men to touch these, that I might be interested in the good omens, which I did, for they were quite harmless, and would tamely suffer themselves to be handled; and then they were put into a large open earthen pan, and set on one side of the highway.[22]

Snakes, both venomous and harmless, are common in West Africa and have inspired numerous and diverse beliefs and cultural practices. A description of a snake crowing like a cock is puzzling, however, since no snakes exhibit this behaviour and therefore this story must contain either error or embellishment. The simplest explanation is that Equiano merely misremembers; he may easily have mixed up in his mind separate childhood incidents involving snakes and cocks, conflated memories of seeing a snake with those of hearing a cock crowing nearby, or is recalling some sort of performance or narrative featuring a crowing snake. This may also be true if his childhood was spent in South Carolina, which has many snake species. There is, however, some evidence that the crowing snake, sometimes also described as feathered, is a mythological creature in south and east African folklore, although apparently not in West Africa, where Equiano described spending his childhood. If the Carolina birthplace hypothesis is correct, this story may have been the adult Equiano incorporating into a fictionalized account of his childhood a folk story that he had heard from other enslaved people in America or the Caribbean. As is so often the case with the opening chapters of *The Interesting Narrative,* we shall probably never know. Nevertheless, the crowing snake aside, these and other passages describing the African environment in Equiano's *Interesting Narrative* do align the text with Benezet and Clarkson's representations of Africa as home to a diverse fauna. The effect for European readers is to augment both the familiarity of the text and the authority of the author—both important considerations for an African writer describing landscapes and societies that would have been exotic to most of his British readers.[23]

Equiano's depictions of wildlife beyond Africa are mostly maritime and are marked by his sense that the ocean is not merely unfamiliar but unnatural:

"Not being used to the water," he tells us on being put aboard a slave ship, "I naturally feared that element the first time I saw it." Equiano often articulates a fear that the animals he encountered might eat him, or, as several critics have noted, displaces what was apparently a deep-seated fear of cannibalism onto these animals. A frequently examined example comes in the *Narrative*'s third chapter when the young Equiano crosses the Atlantic from west to east for the first time. The passage took longer than usual and provisions ran low. "In our extremities," recalls the older Equiano:

> the captain and people told me in jest they would kill and eat me; but I thought them in earnest, and was depressed beyond measure, expecting every moment to be my last. While I was in this situation one evening they caught, with a good deal of trouble, a large shark, and got it on board. This gladdened my poor heart exceedingly, as I thought it would serve the people to eat instead of their eating me; but very soon, to my astonishment, they cut off a small part of the tail, and tossed the rest over the side. This renewed my consternation; and I did not know what to think of these white people, though I very much feared they would kill and eat me.

The maimed shark, unable to swim, and therefore to breathe, would have suffered a painful death shortly thereafter. The cannibalism trope, as Mark Stein has argued, becomes a recurring feature of Equiano's text through which he "foregrounds the reality that African slaves can still be bodily incorporated by the plantation economy." Stein also notes a related incident shortly after in which Equiano was alarmed by "some very large fish, which I afterwards found were called grampusses." These animals, which Equiano believed "to be the rulers of the sea" with the power to control the wind and waves were probably killer whales (*Orcinus orca*), which are in fact mammals, not fish. Equiano recounts his fear that he would be "offered up to appease them" by the mariners. Instead, he was reassured by another child on board, Dick, who would become his close friend—and arguably, as Vincent Woodard has suggested, the object of his "romantic love." These scenes offer a complex interplay of rhetorical structures, driven by the tension between the experienced author and his implied reader on the one hand and the naïve child narrator on the other. The deliberate strategy draws attention to the recently enslaved child's fear and vulnerability, but also emphasizes his profound change in environment, from a childhood in a landlocked African village to what, as it would turn out, would be a life spent largely at sea. In *The Interesting Narrative,* these unexpected, unwished for, and disturbing shifts in natural environment allow Equiano to articulate the unfamiliarity and danger of his new existence, but also become an extended metaphor for the forced migration of enslaved peoples.[24]

In adulthood, Equiano apparently retained a fear of cetaceans, as when sailing along the Maryland and Delaware shore in 1766 he recounts being "surprised at the sight of some whales, having never seen any such large sea monsters before," but when the animal-cannibal conflation resurfaces in adult life it is in relation to birds, not whales. Shipwrecked in the Bahamas the following year, Equiano notes "some very large birds, called flamingoes: these, from the reflection of the sun, appeared to us at a little distance as large as men; and, when they walked backwards and forwards, we could not conceive what they were: our captain swore they were cannibals." The idea "created a great panic" among the mariners, but this time it is Equiano who deals with the situation rationally. He proposes a direct approach so that " 'perhaps these cannibals may take to the water." Accordingly, "we steered towards them; and when we approached them, to our very great joy and no less wonder, they walked off one after the other very deliberately; and at last they took flight and relieved us entirely from our fears." The anecdote reveals the nervousness of the mariners—the American flamingo (*Phoenicopterus ruber*) grows to a height of around 1. 4 meters (4 feet, 7 inches), which is short for a man even by eighteenth-century standards—but it also serves to establish Equiano's growing credentials as a competent seaman and decisive leader. It is one of a series of incidents in the autobiography in which he closes the gap between the naïve child and the authoritative abolitionist, in this case by showing him overcoming his childhood fear of cannibalism and demonstrating a more secure understanding of the natural world.[25]

Equiano's engagement with nature, then, operates in two distinct ways to establish his own authority to enter the debate about the slave trade and then to actually participate in that debate. Like Ramsay and Clarkson before him, he frames much of his discussion of African society in terms set up by John Green in *Astley's Collection,* mediated through Anthony Benezet's *Some Historical Account of Guinea,* while his vision of a diverse and productive African environment again owes much to *Astley's Collection,* mediated both by Benezet and Clarkson. Equiano's decision to rest much of his description of Africa firmly on Benezet is not in itself proof that he made up an African childhood. Instead, it shows that he was deliberately and strategically locating his testimony among the central texts of the abolitionist movement and the sources from which they drew. In accessing the archive, Equiano was no different from Ramsay, Clarkson, and many abolitionists. His descriptions of encounters with sharks, whales, flamingoes, and even crowing cocks offer a more individual perspective but play an important part in driving forward a narrative of personal development in the face of extraordinary circumstances. Both strategies contribute to Equiano's authority in the debate and both are inextricably entangled with abolitionism's long engagement with nature and natural history.

The three authors examined in this chapter each made crucial and original contributions to the abolition debate: Ramsay with his firsthand account of plantation slavery, cast in part in the form of an alternative plantation management manual; Clarkson with his meticulous academic investigation into slavery and the slave trade, in part delivered as an imaginative reconstruction of African society based on Benezet and others; and Equiano with his personal account of the experience of enslavement and self-emancipation. Sharing much of the same source material, and building on each other's work, the three authors reached similar conclusions. They agreed that British colonial slavery was cruel, contrary to scripture, and of dubious legality, but they also saw the slave trade as an impediment to fair trade in Africa's prodigal natural assets and plantation slavery as mismanagement of Caribbean agricultural resources. The slavery system was not merely unnatural, in their analysis, it also obstructed the proper use of God's providence as it was manifested in the natural world. While Anthony Benezet's work can be seen as the prism through which earlier environmental writing was passed and refracted into abolitionism, reaching for the naturalist archive either directly or via Benezet became an abolitionist habit of thought by the 1780s. By this time, the proliferating corpus of abolitionist writing was deeply intertextual and, while abolitionist writers openly acknowledged their debt to many sources and genres, including the tradition of agricultural and natural history writing, peppering their pamphlets and essays with footnotes, quotations, and allusions to the writing of explorers and naturalists from across the centuries offered them a ready method for demonstrating that slavery was an unnatural practice.

CHAPTER 7

"But say, whence first th'unnatural trade arose?"

ABOLITIONISM'S ENVIRONMENTAL POETICS

Antislavery opinion had been voiced throughout the eighteenth century, but in the 1780s it went mainstream. From the middle of the decade onward, abolitionists and their supporters published hundreds of pamphlets, essays, treatises, and sermons urging an immediate end to the slave trade. These appeared alongside antislavery poems, plays, and novels of variable length and quality, often expressed in the then fashionable language of sentiment and sensibility. Many antislavery publications accordingly focused on the physical and emotional suffering caused by slavery and the slave trade. While the human cost of slavery was uppermost in most abolitionist authors' minds, many were also strongly influenced by the tradition of representing colonial slavery in natural histories and agricultural writing, particularly as refracted through the lens of Anthony Benezet's and Thomas Clarkson's antislavery works, which between them offered plentiful source material for creative writers. In this abolitionist literature, Africa was frequently represented as a land of prodigious natural resources despoiled by the slave trade, and the Caribbean as the location of plantations that pushed both people and the environment beyond their natural limits. This chapter explores these representations in a selection of popular antislavery poems from the 1770s and 1780s. All abolitionist literature was "popular" in the sense that it sought a large audience in the hope of effecting political change, even though the meaning of "popular culture" is somewhat different when applied to an age when books might cost many multiples of an average daily wage and when advanced literacy was confined to a minority. Nevertheless, beyond the high-minded nonfiction treatise, pamphlet, and sermon, we can also speak of an abolitionist popular culture that was manifested primarily in plays, novels, and poems. Poets in particular were able to respond quickly to the demands of both the literary marketplace and the news media since poetry had the advantage of being quicker to produce and cheaper to reproduce than novels or dramas. As John Oldfield has noted, the committee of the Society for Effecting the Abolition of the Slave Trade (SEAST) "were not 'saints' but practical men who understood about the market and about consumer choice." Poetry provided an excellent opportunity to bridge the gap between activists and the public, and many who supported the abolition of the slave trade did so primarily through their literary choices. The poetry of antislavery aimed at a wide readership and was apparently consumed by a broad cross-section of the reading public. Quantifying how broad the readership was is difficult, although William Wordsworth's (1770–1850) description of the abolition movement as "a whole country crying with one voice" gives some indication of the scale, the unity, and the emotional tenor of the literature. In *Amazing Grace,* his noted anthology of poems about slavery, James Basker includes "some 400 different titles by more than 250 different writers," most of them opposed to slavery. Many others could have been included, particularly if he had incorporated more from newspapers and periodicals as well as poems written in languages other than English. Nevertheless, by any standards abolitionist verse was extensive and popular, contributing substantially to the public debate that became one of the principal topics of conversation in 1780s Britain.[1]

The poems discussed here range from 1773 to 1788 and each was, in its own way, an original and substantial contribution to the public debate. *The Dying Negro,* coauthored by John Bicknell (1746–87) and Thomas Day (1748–89) between 1773 and 1775, responded to the recent Mansfield decision in the Somerset case with the intention of moving the discussion on to a broader conversation about the slave trade. *West-Indian Eclogues,* written by Edward Rushton (1756–1814) in 1787 offered a response to James Grainger's *The Sugar-Cane* that explored the psychological as well as the physical torments of plantation slavery. In *The Wrongs of Africa, Slavery, a Poem,* and "The Negro's Complaint," William Roscoe (1753–1831), Hannah More (1745–1833), and William Cowper (1731–1800), respectively, contributed poems directly to the SEAST committee to promote the cause. What all these poems had in common was a conception of slavery as unnatural or, at the very least, at odds with the natural environment. To make that case, most drew either explicitly or implicitly on the archive of writing about slavery in natural histories of Africa and the Caribbean. This is not to say that these poets adhere meticulously to the information they found. The poetry sometimes cites the archive, sometimes alludes to it, but is never narrowly constrained by it. Creative writing has a different relationship with fact than polemic, and poets create imaginative scenarios inspired by the archive but not necessarily based on it. How far poetry changed the minds of anyone reading it is debatable, and these poems probably did more to rally the cause than recruit to it. Nevertheless, the poetry of antislavery offers an important insight into the abolitionist mindset at its moment of peak popularity and confirms that the view that slavery was unnatural was widespread and common.

THE DYING NEGRO

The Dying Negro, coauthored in three incrementally expanding editions between 1773 and 1775 by John Bicknell and Thomas Day, is a piece of narrative verse consisting of 434 lines of rhyming couplets in its final version, with the fifth edition of 1793 identifying precisely which lines were Bicknell's and which were Day's. It tells the story of an unnamed "Negro" in London who, although engaged to be married to a white English woman, is against his will put on board a ship bound for the West Indies. Before the ship can sail, however, he takes a knife and stabs himself to death. Much of the poetry consists of interior monologue as the "Dying Negro" reflects on his own situation or the problem of slavery more generally, but while this aspect of the poem was drawn from the poets' own imaginations, the narrative was inspired by an actual story reported in several London newspapers in May 1773: "Tuesday a Black, Servant to Capt. Ordington, who a few days before ran away from his Master and got himself christened, with the intent to marry his fellow-servant, a White woman, being taken and sent on board the Captain's ship in the Thames, took an opportunity of shooting himself through the head." While this brief but shocking news report was the immediate spur, the poem can also be considered a response to the celebrated Mansfield ruling in the James Somerset case, made less than a year earlier, since the newspaper reports that inspired *The Dying Negro* showed that Captain Ordington was in breach of the law as it now stood. The poem was thus a localized protest against Ordington and a call for proper enforcement of the Mansfield ruling as much as it was a more generalized critique of slavery.[2]

Although cowritten, the poem is often discussed solely as the work of its most celebrated author, Thomas Day, even though Bicknell appears to have supplied the first draft, which Day then polished and augmented. The two were young lawyers in 1773, and had no doubt followed the Somerset case closely. Day is better remembered not as a lawyer but as a devotee of Jean-Jacques Rousseau (1712–78), as one of the members of the Lunar Society, as a pioneering animal rights activist, as an agricultural reformer, as a children's novelist, and, somewhat less sympathetically, as the man who tried to educate two orphan girls on Rousseauean principles in the hope that one of them would make him a good wife. Neither did, although one, Sabrina Sidney (1757–1843), later married Bicknell, which ruptured the friendship between him and his coauthor. Bicknell himself went on to have a less glorious career. He wrote no further poetry, neglected his legal practice, drank heavily, and died of a stroke at the age of forty-one.[3]

The Dying Negro, although set in the hold of a ship at anchor in what was then the world's busiest port and largest city, is both rich in natural imagery and replete with references to slavery and Africa directly inspired by Anthony Benezet's *Some Historical Account of Guinea.* Bicknell's lines are presented without an apparatus, but Day backs up his with a small number of footnotes citing reliable sources. These refer to three texts. The lines "Nine days we feasted on the Gambian strand, / And songs of friendship echo'd o'er the land" are backed up by a quotation from "*M. Adanson's voyage to Senegal, &c.*," while the lines "I woke to bondage and ignoble pains, / And all the horrors of a life in chains" are supported by two quotations, one from "*Smith's Voyage to Guinea*" and the other from "*J. Barbot's Description of Guinea.*" In fact, Day was being disingenuous. He had not read widely in African natural history. Minor variations in typography and orthography show that his quotations are all taken directly from Benezet's *Some Historical Account,* which had been published in London the previous year. Clearly, Day had both the morning papers and Benezet on his desk in late May 1773. The first of Day's footnotes offers a verbatim extract of Michel Adanson's vision of Senegal as a natural paradise, as quoted by Benezet, in which Adanson's recalls that an African village had "revived in my mind the idea of our first parents and I seemed to contemplate the world in its primitive state." As we have seen, this passage in Benezet's hands became a powerful piece of evidence for the innate virtue of Africans, African society, and the African environment. In Day's hands, Adanson's paradise is lost, as European "traitors" abuse the hospitality of the virtuous but innocent African villagers whose "songs of friendship echo'd o'er the land" while they entertained the visitors with nine days of feasting on the beach. The leader of these "European robbers" is portrayed as deceptively attractive, whose "golden hair play'd round his ruddy face" while he approaches the village "with insidious smile and lifted hand." This echoes the long tradition that depicted Satan in the Garden of Eden as both beautiful and mendacious, inspired by, for example, the biblical depiction of the satanic king of Tyre in Ezekiel 28, but more recently articulated by poets such as John Milton (1608–1674) and Alexander Pope. Milton's Satan appears "with burnisht Neck of verdant Gold" and "pleasing was his shape, / And lovely," while Pope pithily comments that "*Eve*'s Tempter thus the Rabbins have exprest, / A Cherub's face, a Reptile all the rest." Pope is taking a shot at a political figure of his day, but Milton's representation of Eve in Eden, in which she is shown surrounded by the diversity and beauty of God's creation, is more conducive to a narrative that stresses the damage that slave trading does to nature as well as humanity. The natural freedom enjoyed by the African villagers is destroyed by the slave traders' fraudulent invitation to accept the fruits of European economic and technological prowess when they ask the villagers to "Ascend our ships, their treasures are your own, / And taste the produce of a world unknown." Whereas the villagers had taken "the wand'rers to the freshest spring," instead "The smiling traitors with insidious care, / The goblet proffer." The natural products of the environment, water and fruit, are (in the eyes of the teetotal Day) perverted into the unnatural substance wine. The villagers awake from the unexpected drunken slumber to find themselves betrayed into "bondage and ignoble pains" and, like Adam and Eve, are forever exiled from their native land.[4]

While Day is playing on the perennial theme of the fall from paradise in these lines, he is primarily able to do so because of the opportunity offered by Benezet's quotation of Adanson's reference to "our first parents." Day's reliance on Benezet places the poem, therefore, in the tradition of abolitionist literature that was underpinned by natural history. Scholarly discussion of the poem has tended to focus on its sentimental depiction of suffering and doomed love on the one hand or its representation of the grim reality of suicide among enslaved Africans on the other. These are both important elements of what is a complex piece of writing, but the poem is also notable for repeatedly contrasting imagery of the natural world as a place of apparent freedom with the unnatural practices and locations of slavery. This poetic strategy becomes evident early on when the eponymous "Negro" of the poem looks out toward the shore of the Thames from the ship's cabin in which he is imprisoned:

Ye waving groves, which from this cell I view!
Ye meads now glitt'ring with the morning dew!
Ye flowers, which blush on yonder hated shore,
That at my baneful step shall fade no more,
A long farewel!—I ask no vernal bloom—
No pageant wreaths to wither on my tomb.
—Let serpents hiss and night-shade blacken there,
To mark the friendless victim of despair!

The rural scene is not entirely fanciful. Groves of trees and dew-covered flower meadows could be seen from oceangoing ships on the river as little as 2 kilometers (1.2 miles) east of London Bridge, although keen-eyed observers would have to have looked over the narrow ribbon of shipwrights, timber yards, and dry docks that lined the shore from London to Greenwich and beyond. The passage neatly upsets the delightful landscape, however, turning flower meadows into plantations for funeral wreaths, which are then explicitly rejected in favor of snakes and nightshade—presumably deadly nightshade (*Atropa belladonna*)—a plant long used to produce a poison effective both for murder and suicide. While the cabin is the immediate scene of captivity, the landscape itself foreshadows the suicide to come, but also perhaps reminds the reader that the managed landscapes of groves and meadows are the temperate equivalent of the tropical plantations that were deadly for so many Africans. In any case, this "hated shore" is England and the ship's destination is the Caribbean. Thinking ahead to the vessel's geographical terminus, the "Dying Negro" imagines a typical "day of misery" endured by enslaved laborers on a West Indian sugar plantation. The day ends with their sleep, which he describes, as if speaking directly to his fellow Africans on the other side of the Atlantic, as "the only boon of heav'n / To you in common with your tyrants giv'n." The consolation of sleep is to dream:

In sweet oblivion lull awhile your woes,
And brightest visions gladden the repose!
Let fancy then, unconscious of the change,
Thro' our own fields, and native forests range;
Waft ye to each once-haunted stream and grove,
And visit ev'ry long-lost scene ye love!

If in their dreams the enslaved people of the plantation return to Africa, it is significant that they return to their fields, forests, streams, and groves rather than to their towns, villages, homes, and families. In the poem's scheme, Africa is at least partially cultivated—it too has fields and groves—but it is also very largely a wild place of forests and streams in which the youthful narrator once hunted ferocious animals such as "The howling tyger, and the lion grim," unaware that the most dangerous of all were "human brutes more fell, more cruel far than they." The image of the young African man as a hunter of wild animals just prior to being enslaved by Europeans was fast becoming a poetic convention at this time. It had appeared, for example, in Thomas Chatterton's (1752–70) short poem "Heccar and Gaira: An African Eclogue," published a few years earlier in 1770. This poem may have influenced Day and Bicknell, not least because Chatterton's own suicide had been a matter of public sensation. *The Dying Negro* is, however, a more expansive work than "Heccar and Gaira" and it shifts not only in place but also in time: from the narrator's past to his present and then implicitly to the future moment in which he takes his life. A repeated theme throughout the poem is the comparison between sleep and death, which to some extent echoes Hamlet's suicidal thoughts—references in the poem to several plays confirm that Bicknell and Day were keen Shakespeare fans. It is also a reference to the then well-known idea that enslaved Africans in the Caribbean believed they returned to Africa on their death, which is alluded to several times in the poem, with the implication that the overturned natural order is put back into its original state after death. This sense that European empires have turned the world upside down is likewise returned to frequently, both directly and implicitly. As the sun rises, the "Dying Negro" asserts that, for him, it means darkness:

> Yon ruddy streaks the rising sun proclaim,
> That never more shall beam upon my shame;
> Bright orb! for others let thy glory shine,
> Mature the golden grain and purple vine,
> While fetter'd Afric still for Europe toils,
> And Nature's plund'rers riot on her spoils;
> Be theirs the gifts thy partial rays supply,
> Be mine the gloomy privilege to die.

Again, the personal is conflated with the natural and the rural. While the sun figuratively sets on the despairing narrator, it literally rises over and nourishes an empire built on partiality and pillage. Bicknell and Day cast Europe's colonists in the role of plunderers, but while "fetter'd Afric" toils for Europe, it is nature that is plundered. That this is a violation of the natural order is again reinforced in lines that characterize slave traders as "hell-hounds" but that reassert the common humanity of Europeans and Africans despite their different environments:

> Where-e'er the hell-hounds mark their bloody way,
> Still nature groans, and man becomes their prey.
> In the wild wastes of Afric's sandy plain,
> Where roars the lion thro' his drear domain,
> To curb the savage monarch in the chace,
> There too Heav'n planted Man's majestic race.

The argument that God "hath made of one blood all nations of men for to dwell on all the face of the earth" (Acts 17:26) taken from Paul's Aeropagus sermon, was frequently cited in antislavery literature. The scripture verse concludes that God "hath determined the times before appointed, and the bounds of their habitation" and, although Bicknell and Day do not quote chapter and verse, they certainly invoke it. Throughout the poem, if nature is being violated, they are clear that this is the nature that God had created and within which God had put everything in its proper place. Indeed, God is figured in this final line of this extract as a planter, who alone has the authority to decide where "Man's majestic race" is cultivated. The forced mass transplantation of people from Africa to America and the Caribbean in this analysis is not only an assault on nature but also a violation of God's will.[5]

The contrast between "fetter'd Afric" and unfettered nature is repeatedly made throughout *The Dying Negro.* Nowhere is this more apparent than in an extended passage, about two-thirds of the way through, in which the narrator addresses his absent lover and imagines what he and she could do were he able to free himself:

> O could I burst these fetters which restrain
> My struggling limbs, and waft thee o'er the main,
> To some far distant shore, where Ocean roars
> In horrid tempests round the gloomy shores;
> To some wild mountain's solitary shade,
> Where never European faith betray'd.

This distant shore, free from European contact, is explicitly not Africa. The poem hints that it is a polar region (whether north or south is not specified), but it is quite obviously a place in the realm of the imagination that the narrator will never see. In the lines that immediately follow, the narrator imagines re-creating himself in this place as a man of the wilds, akin to the archetypal "noble savage" of the European imagination, but in some ways more like Crusoe on his island. To protect his love on these "gloomy shores" he would "meet ev'ry danger," would "climb the rock, explore the flood, / And tame the famish'd savage of the wood." It does not appear to trouble the poem's authors that his taming of the "famish'd savage" implicitly makes the enslaved narrator a colonist himself. Such contradictions may be ascribed to poetic license, but they also suggest that the idea of expansion and colonization was deeply embedded, even in the minds of those who critiqued it. Indeed, the narrator goes on to offer a vision of himself as a tamer not only of savages but of harsh and alien environments:

> Nor snows nor raging winds should damp my soul,
> Nor such a night as shrowds the dusky pole;
> O'er the dark waves my bounding skiff I'd guide,
> To pierce each mightier monster of the tide;
> Thro' frozen forests force my dreadful way,
> In their own dens to rouze the beasts of prey;
> Nor other blesing ask, if this might prove
> How fix'd my passion, and how fond my love.

Thus, the narrator, the "Dying Negro" who is a victim of European colonization, unfettered trade, and environmental exploitation becomes himself an exploiter and colonizer of continents beyond Europe, America, or Africa. Since this is in the cause of true love rather than imperial trade, however, the reader is invited to sympathize with it. The poem may be somewhat muddled in its geopolitics, but its depiction of the suffering of an individual victim of a slave trade that was as unnatural as it was cruel was in its time an effective strategy.[6]

In earlier editions, the poem's closing lines saw the narrator bitterly attack and then reject the Christian God: "When crimes like these thy injur'd pow'r prophane," he exclaims, "O God of Nature! art thou call'd in vain?" The invocation of God as "God of Nature" is consistent with the poem's theme of natural freedom versus unnatural slavery, but in the first edition, God is rejected nonetheless. In later editions, however, the narrator receives indications that God will facilitate his revenge:

> —Thanks, righteous God!—Revenge shall yet be mine;
> Yon flashing lightning gave the dreadful sign.

I see the flames of heav'nly anger hurl'd,
I hear your thunders shake a guilty world.
The time has come, the fated hour is nigh,
When guiltless blood shall penetrate the sky.

What follows is a series of "prophetic visions" in which "eternal justice wakes" and unleashes the final apocalypse. In forty lines of lurid verse, the "plagues of Hell" are unleashed on Europe, in particular its system of international trade. Discord "fires all her snakes" until "No more proud Commerce courts the western gales, / But marks the lurid skies, and furls her sails." Next, War ravages the continent until "with horror sick'ning Nature groans," Europe's "flaming cities crash around," and "In mournful silence, desolation reigns." Finally, when Armageddon has been unleashed, at least in the Northern Hemisphere, "Afric triumphs." It is tempting to read this as a purely environmental catastrophe brought about by the despoliations of the slave trade, but it is also quite clearly in the Christian millenarian tradition, albeit with Africa spared the general devastation. Nevertheless, the poem clearly and repeatedly posits slavery and the slave trade both as an assault on God and as a series of depredations on the natural world that God created. While the occasion of the poem is the failure of the authorities to enforce the 1772 Mansfield decision, its primary sources come via Benezet's *Some Historical Account of Guinea.* As one of the most powerful and most widely read statements of antislavery of the 1770s, this two-pronged attack would go on to be highly influential. It is clear that *The Dying Negro* did much not only to keep alive the growing public sense of the injustice of slavery that had been awakened at the time of the Somerset case but also to promote Benezet's conception of Africa as a place of prodigious natural and human resources despoiled by the unnatural slave trade.[7]

EDWARD RUSHTON'S WEST-INDIAN ECLOGUES

In late 1787, more than a decade after Bicknell and Day's *Dying Negro* and following the formation of SEAST in May, the blind Liverpool poet and former slave-ship seaman Edward Rushton (1756–1814) published a work called *West-Indian Eclogues* that included three verse dialogues between enslaved Africans on a Jamaican plantation plus a similar poetic monologue. The poems are notable for their unsparing depiction of the cruelties of slavery from a laboring-class poet who had witnessed the realities of plantation life firsthand, but also for their substantial and explicit engagement with Caribbean natural history, both in their form, diction, and setting and in their extensive biological and ethnographic endnotes. By locating the cruelties of slavery in the Caribbean environment, and by realizing that environment through the natural history archive, Rushton's eclogues powerfully assert that slavery is unnatural as well as cruel.

Edward Rushton was born into a modest family in 1756 in Liverpool, at around the time when it was overtaking Bristol as England's largest slave-trading port. At the age of ten he went to sea and later joined slave-trading voyages to West Africa and the Caribbean. At just seventeen, he was blinded by an eye infection contracted on board a slave ship near Dominica. Returning to Liverpool, he initially lived in poverty before being set up as a tavern keeper by his father, but his talents ran more in a literary direction and he published verse, briefly edited a local newspaper, and eventually became a bookseller. He made no secret of his radical political views, which affected his trade, but he was eventually able to secure enough business to lead a comfortable life. He voluntarily underwent a horrendous operation on his eyes in 1807, which restored the sight in one eye. He died in 1814. His poetry is marked by its political radicalism on the one hand and its attention to the natural world on the other. While much of the visual imagery of his poetry recalls his life at sea, many of his poems celebrate birds through their songs and calls, revealing Rushton both as a skillful bird listener and a sensitive nature poet. His own blindness forms the subject of several of his poems, but the major theme running throughout his corpus is resistance to oppression, of which slavery is the most egregious example. *West-Indian Eclogues* is the most substantial contribution to this body of work.[8]

West-Indian Eclogues is divided into six distinct parts. It begins with a conventionally modest foreword followed by a short dedication to Beilby Porteus (1731–1809), the Anglican bishop who had delivered an important sermon on slavery in 1783. The prefatory material aligns the poem both with the prevailing ethos of literary abolitionism, which often adopted a rhetoric of sensibility to assert its humanitarian principles, and with a substantial figure of the English establishment. These are tactical choices from a radical author seeking to make an impact on mainstream politics. The following sections constitute the poem itself, divided into four eclogues—essentially poetic dramatizations of fragments of time that illustrate broader themes. These include three dialogues that take place between enslaved people on a plantation over the course of two days, plus one monologue. In the first, Adoma reveals to his friend Jumba one morning that he had been whipped for trying to protect his wife from a brutal overseer while she was breastfeeding their son, and the two plot revenge. In the second eclogue, set later in the day, Adoma reveals to Jumba that he has been having doubts about the plan; Jumba attempts to persuade him to take action and they go their separate ways. In the third, set the following morning, Quamina tells Congo that he just saw Jumba "number'd with the dead" and recounts the tale of "an ancient Slave," Angola, who though too old to work was beaten to death "beneath the driver's lash." The final eclogue is a monologue from Loango, who is driven to violent revenge when his wife Quamva is raped by an overseer.

The poems end with Loango's revenge, but the book as a whole concludes with an extensive set of explanatory and discursive endnotes. Given that the introductory material and the notes weigh in at around 3,200 words, while the poem itself is only a little longer, at 3,800 words, the apparatus is clearly as important as the text itself, even if critics have not always been certain what to make of it. Joshua Crandall, for example, observes a strong "distinction between the rhetoric and tone of Rushton's notes and the bodies of his poems." While the poems contain "highly sentimental and sympathetic depictions of the plight of African slaves," he argues, the endnotes "emphasize authoritative observation." Franca Dellarosa finds the notes a "somewhat perplexing counterpart" to the poetic text, especially where the radical calls to violent action articulated by the enslaved narrators of the poem are apparently undercut by a more conservative appeal for plantation amelioration in the notes, but she concludes that their function is ultimately "to endorse the authenticity of the poetic narrative and dramatic characters with factual substance." This is certainly true, but the notes also align the text as a whole with the tradition of representing slavery in environmental writing, even though Rushton names only one scientific and one abolitionist source: the former an account of the flying fish in *The Philosophical Transactions of the Royal Society* and the latter "The Rev. Mr. RAMSAY's Treatise." Other influences are, however, apparent. "In the third eclogue, for example, Quamina reminisces about his life in Africa, recalling "times, when we chac'd the fierce-ey'd beasts of prey" and "oft we saw, in spite of all his care, / The bulky Elephant within our snare." The supporting footnote offers an account of trapping elephant in a pit that is similar to that given by André Brue in *Astley's Collection.* Rushton's account of the atrocious torture in which enslaved people were whipped and "pepper, and salt, are frequently thrown on the wounds, and a large stick of sealing-wax dropped down, in flames, leisurely upon them" derives ultimately from Hans Sloane's *Voyage to Jamaica,* but the wording reproduces exactly the version given by Anthony Benezet in *Some Historical Account of Guinea.* These and other examples show that the *Eclogues* and their endnotes were informed by wide reading in the archive.[9]

Critical responses to Rushton's *Eclogues* have been mixed, although it is fair to say that modern readers of the poems have found more to praise than did Rushton's contemporaries. The *English Review* was succinct but enthusiastic, arguing that "the several eclogues are written in a natural strain of poetry, beautifully enriched with local images and allusions, and highly descriptive of the characters and sentiments of the sable race of mankind. They discover such a degree of genius as we have never before seen employed on the subject of slavery." Such high praise was not matched by other reviewers. Andrew Becket (1749–1843), writing in the *Monthly Review,* felt that the versification was "not unpleasing" but otherwise used the review to claim that "writers have greatly exaggerated in their account of the cruelties exercised towards the Negroes" and to assert that such violence was "a kind of political necessity." The *Critical Review* felt that "these eclogues appear more commendable for their design than execution." The reviewer was troubled by the poem's representation of enslaved people resisting their enslavement, arguing that "tears and supplications, not the impotent rage and defiance of the wretched, are most likely to melt their persecutors' hearts." The demand for a less radical and more sentimental antislavery rhetoric represented mainstream abolitionism in this period since, whatever they might have thought in private, few abolitionists wanted to publicly endorse violence against British colonists. As well as his critique of Rushton's radicalism, however, the reviewer also betrays his disdain for Rushton's laboring-class origins through his commentary on the poet's use of nautical language. In a note on the line "Rising from distant reefs and rocky shores," he asks, "Are not *reefs* and 'rocky shores' synonymous? The word itself, we apprehend, . . . is merely a nautical one, and not to be found in any dictionary." Although the two are often found close together, a reef may exist separately from a shore, as any mariner would know. The comment reveals the reviewer's snobbery as much as his ignorance. While perhaps sympathetic to the abolitionist cause, he clearly does not believe that writing antislavery verse is the business of laborers, either free or enslaved.[10]

By contrast, recent critics have admired the poems' radicalism. Stuart Curran pithily describes them as depicting "the mercilessly oppressed exploding in murderous revenge." Dellarosa makes a case "for firmly aligning *West Indian Eclogues* within Rushton's most radical poetic *and* political testimony."

Grégory Pierrot argues that Rushton was "a poet of the 'Atlantic proletariat' and studying his texts reveals something of the complex and untidy ways in which capitalism and slavery intersected in eighteenth-century British writings." Several critics have commented on the poet's use of form. Dellarosa argues that "Rushton carried out a sustained experiment in a form whose inherent dramatic power had proved to be particularly apt to respond to many late eighteenth-century concerns." In particular, she points to Chatterton's "Heccar and Gaira" and a series of antislavery "American Eclogues" published by Rushton's friend Hugh Mulligan in the *Gentleman's Magazine* in the early 1780s as inspiration for the poem's form. It was the eclogue's "quasi-dramatic structure," she argues, that allowed it to locate the poem at a specific time and in a specific location while giving a voice to those who had been otherwise silenced. Rushton was indeed a fan of Chatterton and wrote a poem in his memory called *Neglected Genius: or, Tributary Stanzas to the Memory of the Unfortunate Chatterton,* which was apparently published at the same time as *West-Indian Eclogues.* As well as the form, his *West-Indian Eclogues* share with Chatterton's *African Eclogues* a pithy and direct style and a willingness to countenance violent resistance to enslavement.[11]

Antislavery poems by Mulligan and Chatterton may not have been the only influence. In a brief note, Rushton's modern editor Paul Baines observes that "Rushton was also parodying the colonial georgic of James Grainger's *The Sugar-Cane.*" Baines does not elaborate further, but the observation is precise. The first clue lies with Rushton's choice of the eclogue form, which pits the poem against Grainger's georgic in that the *Eclogues* and the *Georgics* constitute the two halves of Virgil's pastoral output. Both Grainger and Rushton transcribe the scene of rural labor from an Italian farm to a Caribbean plantation but, where Virgil's *Georgics* described the operation of the farm from the point of view of the landowner, the first and ninth of the ten *Eclogues* deal with, among other things, enslavement and agrarian displacement and are generally from the perspective of laborers rather than landowners. The contrast would have been immediately apparent to most late eighteenth-century readers, who may also have noticed that Rushton offers four eclogues to match the four books of Virgil's *Georgics* and Grainger's *Sugar-Cane* as well as a set of natural history notes almost as extensive as Grainger's.[12]

Each of the *West-Indian Eclogues* falls into two parts: a locodescriptive opening followed by a dialogue or monologue. The dramatic parts have interested critics more than the descriptive openings. These powerful human stories about abuse, resistance, and revenge, while radical in suggesting that violence might be a legitimate response to enslavement, clearly align with abolitionist discourses that sought to expose the cruelty of plantation slavery through representations of suffering enslaved Africans. The locodescriptive openings are by contrast often considered as little more than "framing" sequences to the human stories. As the poems develop, however, the distinction between the human and the natural world is progressively blurred, and the enslaved protagonists are increasingly equated both with nature and with natural justice. By the end, the frame and the content are merged as the eclogues reach their dénouement. The first eclogue, however, does indeed concentrate on scene setting, with a subtitle announcing "SCENE—JAMAICA.—TIME—MORNING." The first six lines establish the location as a tropical island at dawn:

> THE Eastern clouds declare the coming day,
> The din of reptiles slowly dies away.
> The mountain-tops just glimmer on the eye,
> And from their bulky sides the breezes fly.
> The Ocean's margin beats the varied strand,
> It's hoarse, deep, murmurs reach the distant land.

Except for the single reference to "the din of reptiles," the scene could be any location where there are mountains close to the shore, as there are in several places in England, Wales, and Scotland. Rushton's observation about reptiles, which were probably in fact amphibian tree frogs (Hylidae) and rain frogs (Eleutherodactylidae), is supported with an endnote that both elucidates and emphasizes the detail's importance as the key indicator of exoticism in the passage. The scene is, however, remarkably peaceful, idyllic even, once the amphibian "din" has faded away, counterpointing the violence of the coming scenes, with only the single word *beats* hinting at what is to come. Almost immediately, the poetic narrative shifts to Jumba and Adoma, the "sons of Mis'ry" whose story occupies the first two eclogues. The rest of the first eclogue contains barely a further reference to their physical environment, focusing instead on the abusive social relationships of the plantation.[13]

The pattern begins to shift in the second eclogue. The locodescriptive opening is considerably expanded to fourteen lines before Jumba and Adoma come on the scene, while the idyllic morning gives way to a nightmarish evening in which the natural world is, by turns, monstrous, diseased, or discordant. Instead of eastern clouds, unfamiliar tropical stars "shine with more than European light." The breeze of the first eclogue returns, but "on it's dewy wings sits pale disease." The frogs are back, but now as "lurking reptiles" that "fill the air with shrill discordant sounds." They are joined by "unnumber'd insects," among which are found strange fireflies alongside which "the keen Mosquito darts his sting." If Jamaica was initially shown as pleasantly exotic, this is now revealed as a cruel deception. In contrast to the first eclogue, Jumba and Adoma also increasingly turn their thoughts to the potentialities of the landscape. Unwilling to take violent revenge, Adoma considers suicide by leaping "from the craggy steep" or eating the poisonous fruit of the manchineel tree. As an alternative, he wonders if it would be possible to escape:

> By seeking shelter on the mountains' heights,
> Where wild hogs dwell, where lofty Cocoas grow,
> And boiling streams of purest waters flow.
> There we might live; for thou with skilful hand
> Canst form the bow, and jav'lin, of our land.
> There we might freely roam, in search of food,
> Up the steep crag, or through the friendly wood.

In this passage, the natural world is posited in contrast to the social world of the plantation as a place of safety and recourse, with the wood itself personified as "friendly." Adoma's dream of returning to a life of free roaming invokes the many representations in literature of Africans as noble hunters, both in numerous accounts in travelers' tales and those in poems such as Chatterton's "Heccar and Gaira," which Rushton's *West-Indian Eclogues* often recalls. The dream is, however, no more real than Chatterton's literary imaginings. Jumba brings Adoma down to earth by reminding him that "The King of all those mountains is our foe." Rushton provides an endnote about the history of the Jamaican Maroon community, explaining that under the treaties agreed in 1739 and 1740, they were required to capture and return fugitive slaves, but Jumba implies that the natural environment is itself confederate in the process of enslavement. "No rocks can cover us, no forests hide," he laments. "Against us ev'n the chatt'ring Birds combine." Again, there is an endnote that corroborates the phenomenon. Jamaican "blackbirds," says Rushton, "at the sight of a human being . . . begin a loud and continual clamour which is heard at a considerable distance. Their noise serves as a guide to the mountain-hunters, who immediately penetrate into that part of the wood, and seize the fugitives." The bird itself is probably not the Jamaican blackbird (*Nesopsar nigerrimus*) but instead either the noisier Jamaican crow (*Corvus jamaicensis*), known locally as the "jabbering crow," or the greater Antillean grackle (*Quiscalus niger*). In more poetic terms, Rushton is offering a species of what critics today recognize as the pathetic fallacy, in which the hostile social environment of Jamaica is mirrored by a hostile physical environment. Adoma falls into despair and, in an extended metaphor, compares his and Jumba's fate to that of the flying fish they saw being hunted while they were on the Middle Passage:

> Ah, JUMBA, worse, much worse our wretched state,
> Thus vex'd, thus harass'd, than that fishes fate,
> Which frequent we beheld when wafted o'er
> The great rough water from our native shore.
> He, as the tyrants of the deep pursu'd,
> Would quit the waves their swiftness to elude,
> And skim in air:—when lo! a bird of prey
> Bends his strong wing, and bears the wretch away!
> No refuge, then, but death—

Rushton supports the passage with a lengthy endnote referring to "the *flying fish* (the *hirundo,* or *mugil alatus,* of some authors, and the *exocœtus volitans* of the Phil. Trans. vol. 68.)" He does not identify the "some authors," although Hans Sloane uses *Mugil alatus* in his description of flying fish, in the same passage where he also cites Richard Ligon and many others and, as we have seen, the predator-prey relationship between flying fish, dolphinfish, and seabird also interested Griffith Hughes. Rushton may have read some of these, but he also had no doubt witnessed the phenomenon himself on a transatlantic voyage. The endnote grounds the poetic text in the archive of transatlantic voyagers, many of whom had described flying fish, but the poem offers a striking new metaphor in which Adoma and Jumba are depicted as the helpless prey of a natural system, betrayed by birds on land and caught between planters and the Maroon "king" much as the flying fish are caught between dolphinfish and seabirds. Far from offering a vision of slavery as unnatural, the second eclogue seems to be arguing that it may be cruel but that it is an inescapable natural phenomenon.[14]

This does not last. The locodescriptive opening of the third eclogue is again doubled, to twenty-four lines as opposed to fourteen in the second eclogue and six in the first. As in the first, the scene is welcoming, even idyllic. Despite the heat of the noonday sun, "the huge mountains charm the roving eye." The birds "sport and flutter," including hummingbirds that "shew their tints where blossoms most abound." Insects are "variegated," lizards merely "speckled," and even the vulture is "useful." In the midday heat, "wearied Negroes to their sheds repair" and reminisce. Quamina reminds Congo of "Times, when we chac'd the fierce-ey'd beasts of prey / Through tangled woods, which scarcely know the day," a scene consistent with eighteenth-century notions of noble savages hunting in a state of nature that Rushton supports with the long endnote about elephant hunting that probably derived from *Astley's Collection.* The passage acts as a textual bridge between the depiction of natural beauty that begins the eclogue and the story of unnatural violence that follows. Reminiscing over, Quamina and Congo discuss the latest news, including the sudden death of Jumba and the abuse of the "ancient Slave, ANGOLA," both events that explicitly illustrate the brutality of the "flinty-hearted Christians" who are "of earth the bane, of manhood the disgrace." Recalling the ending of Bicknell and Day's *Dying Negro,* the third eclogue concludes by emphasizing that the behavior of Europeans is an aberration from human and natural justice rather than the norm and calls for divine retribution. Nature is no longer apparently complicit with injustice, as in the second eclogue. Instead, the guiding spirit of nature, the Christians' "dread Judge, who, they pretend to say, / Rules the whole world with undivided sway" is called upon to restore natural justice.[15]

The final eclogue completes the reversal of the impression given in the earlier eclogues that nature is complicit with slavery. Switching from dialogue to monologue, and considerably blurring the distinction between external environment and internal emotion, it depicts a hurricane making landfall at midnight while Loango rages at the overseer who has raped his wife Quamva. "On this theme," as Grégory Pierrot has pointed out, the poem "is connected to a complex network of racialized representations of rape in English, and more broadly, Western literature." As well as being "rooted in a long tradition of tying the violation of women to collective politics, it also expresses the deep unease with which even the most progressive European abolitionists considered the possibility of Black agency in the New World." While Pierrot shows how Rushton tackled this "conundrum," both in the *Eclogues* and in later poetry, an ecocritical reading of the poem might focus on the environmental feature of the storm. This could be considered as simply another example of the pathetic fallacy, which Pierrot notes, but Rushton is not merely depicting Loango's internal state mirrored in the storm, nor is the hurricane just a conventional shorthand for divine intervention. Instead, Rushton shows Loango consciously attempting to command the hurricane, that is, to take control over the environment in which he is enslaved. At first, he fears the storm may be deterring Quamva from returning to him, calling out, "Hoarse thunder, cease thy roar:—perchance she stays, / Appall'd by thee, thou light'nings fiery blaze." After a while he changes his mind and instead calls upon the hurricane to take the part of a revenging spirit:

> Roar on, fierce tempests:—Spirits of the air
> Who rule the storms, oh! grant my ardent pray'r.
> Assemble all your winds, direct their flight,
> And hurl destruction on each cruel White:—
> Sweep canes, and Mills, and houses to the ground,
> And scatter ruin, pain, and death around:—

Such devastation was not merely hypothetical, of course. As Rushton spells out in an endnote, hurricanes frequently wreaked widespread damage in the Caribbean, and still do, whether or not spirits are involved. For abolitionist readers, Loango's imprecations remain sympathetic as long as they are directed toward planters, but Rushton then has him turn them toward his wife in a series of actions that recall both Othello and Oroonoko at their least attractive. Still invoking nature, Loango is seized with misogynistic jealousy. Believing that women's affection "changes like the Hurricane's fierce wind," or, less violently, flits about "as shifts the Bee, / or long bill'd Humming-bird from tree to tree," he persuades himself that she has fallen in love with her rapist and, while crying out that "my brain's on fire!" calls upon nature to wreak "vengeance on her head" as well:

> —Ye hidden scorpions creep,
> "And with your pois'nous bites invade their sleep;
> "Ye keen CENTIPEDES, oh! crawl around,
> "Ye sharp-tooth'd Snakes, inflict your deadly wound.

If the catastrophic hurricane offers the potential for a general and public devastation, reptiles and invertebrates present the possibility of an intimate revenge. Rushton leaves it open whether Loango genuinely believes he can command animals and the weather, but in reconciling the enslaved person with the natural environment and in pitting hurricanes and centipedes against the plantocracy, Rushton's final eclogue argues that slavery and nature are at odds—with the implication that the former is unnatural. In this scenario, jealousy and revenge, although dreadful, are natural. In the final lines, set apart from Loango's first-person narrative with a dividing line and spoken by the narrator in the third person, the vengeance is realized and the reader is invited to behold "the mangled dead!" The awful scene provides a radical conclusion to the poem but also illustrates Chatterton's influence on Rushton. The final couplet of the poem reads like a continuation of Gaira's final outburst in "Heccar and Gaira" in which Gaira vows to "strew the beaches with the mighty dead" but Chatterton holds back from showing the deed. As Pierrot has put it, in Chatterton's poem "violent retaliation is at the heart of the text, yet it remains out of it: the warriors' vow never does, and never can, come to fruition." It can, however, in Rushton's hands. This becomes apparent when we stitch the two couplets together into a single quatrain, with Rushton's following Chatterton's:

> I'll strew the beaches with the mighty dead
> And tinge the lily of their features red.
> Then to the place, with frenzy fir'd, he fled,
> And the next morn beheld the mangled dead!—

Although the narrative position shifts from first to the third person, Rushton's couplet extends Chatterton's rhyme, meter, his pattern of alliteration, and the sense of his argument to complete the action that Chatterton had merely implied. It also builds on and perpetuates the image of the African young man as a valiant child of nature that Chatterton, Bicknell, and Day had promoted in the 1770s and that was becoming increasingly widespread in the poetry of the 1780s. In this tradition, if the natural state of African society was freedom, then slavery, which disrupted that society and abducted its young men, was unnatural.[16]

THREE ABOLITIONIST POEMS

Once seen as the triumph of a small group of dedicated "saints" driving forward an evenhanded legislative process, British abolitionism is now better understood as a complex and diverse movement that brought together people of otherwise widely differing personal, ideological, and religious backgrounds. It nevertheless remains true that there was an abolitionist "inner circle" consisting not only of members of the SEAST executive committee itself but also those in their wider social and familial circles. This chapter concludes by reading three poems written by poets connected both to SEAST and to one another to show how they are inspired by the archive of natural history representations of slavery, even where they do not explicitly invoke it, and how they share a conception of slavery as unnatural. William Roscoe (1753–1831),

Hannah More (1745–1833), and William Cowper (1731–1800) were connected through Clarkson, Wilberforce, and others of the abolitionist circle even though the three apparently never met. Unlike the work of Bicknell, Day, Rushton, and many other poets of the period, the antislavery poems produced by Roscoe, More, and Cowper in 1787 were written in close cooperation with SEAST, and as such their antislavery work constitutes the closest approximation we have to an "authorized" abolitionist literary canon.

The first of the three poems to appear was "Part the First" of Roscoe's two-part *The Wrongs of Africa,* which was published early in the summer of 1787 not long after the formation of SEAST on 22 May. Roscoe was not part of the committee but, according to Clarkson, the poem was discussed at the third meeting of SEAST on 7 June since Roscoe had offered to donate any profits to the cause. The first part of *Wrongs of Africa* considered only the slave trade within and to the coast of Africa, not the wider question of plantation slavery, so it may not have been coincidence that immediately after considering the poem, "a discussion unexpectedly arose" among the committee members about whether the campaign would focus on "the evil of slavery itself" or merely "the evil of the Slave-trade." They decided on the latter. Whether Roscoe's poem was a direct influence is not recorded, but it remains at least plausible that the poem had an impact on what, in the event, would be a far-reaching decision. Inspired by the support of the abolition committee, Roscoe published the second part of *Wrongs of Africa* the following year, dealing with the Middle Passage and containing some of the poem's most memorable lines in its account of a shipboard insurrection. A third installment, focusing on "the destination of the slaves, and the severities exercised on them in the colonies," was promised but never appeared, perhaps because its topic was not aligned with SEAST's policy of campaigning only against the slave trade.[17]

The poem received mixed reviews. Writing in the *Monthly Review,* William Enfield (1741–97) echoed the widespread abolitionist belief that simply making the facts available about the slave trade would naturally lead to its downfall. "The poet," he argues, "in order to produce the strongest impression on the imagination and feelings of his readers, has only to follow the track of the historian, and clothe plain facts in the dress of simple and easy verse." Roscoe had achieved this, he felt, "with judgment, taste, and genius." The *English Review* concurred that the poet's "versification is, in general, harmonious, varied, and sweet" and its author "appears to be animated by the best of motives." The reviewer objected, however, to the very idea of writing about the "enormous crimes" of the slave trade in verse. The topic should only be tackled "in the stern language of unimpassioned justice" because "he that seeks by the artifice of rhetoric to rouse our feelings upon a subject like this, is immediately despised for the palpable imbecility of his judgment." Recent critics, while recognizing Roscoe's contribution to the cause, have been divided over the quality of the verse. James Basker notes the poem's "moral indignation" and takes the view that Roscoe depicts a "bloody shipboard insurrection . . . as vividly as any writer of the century." Marcus Wood finds the poem "stiff and over-declamatory, but it reflects the profound liberationist philosophy of a brave anti-slavery activist." Wood's assessment reflects what Jessica Moody has called "the memorial cult of William Roscoe" in which Roscoe, in the absence of many other antislavery figures from the slave-trading city, has become "Liverpool's local counterpart to the national martyr-hero, William Wilberforce." Certainly, Roscoe's principled opposition to the slave trade has been more celebrated than his actual poetry.[18]

Like Roscoe, Hannah More was from a slave-trading city, Bristol in her case, but she was far more personally intertwined with the emerging abolition movement than Roscoe. In the summer of 1786, More spent time with her friend Lady Margaret Middleton (c. 1740–92), the wife of Sir Charles Middleton (1726–1813), at their home at Teston in Kent, from where the Middletons played a pivotal role in bringing together the first generation of abolitionists. James Ramsay had stayed in Teston after his return from St. Kitts and soon became the vicar of the parish church, which was in Middleton's gift. Teston also saw regular visits from Clarkson and Wilberforce, both of whom received encouragement and political backing from Middleton. Following her stay in 1786, More would later make the acquaintance of Clarkson and would become a firm friend of Wilberforce, but she also became convinced that poetry was an important weapon in the abolitionist arsenal. *Slavery, a Poem* was accordingly composed in the autumn of 1787 and published on 8 February 1788. In the words of James Basker, the poem is "grounded in Christian faith and pitched as a political manifesto." Framed by an extended address to the spirit of liberty, the poem is also notable for its extended critique of emerging racial ideologies, although in its now infamous contention "Tho' few can reason, all mankind can feel," More reveals her patrician instincts. Contemporary reviews often paid more attention to the context than the text. Writing in the *Monthly Review,* Samuel Badcock (1747–88) contrasted More's poem on the slave trade with that of her protégée, the laboring-class poet Ann Yearsley (1753–1806), which had appeared by coincidence at almost exactly the same time. Of Yearsley, Badcock commented that "in the heat of invective, she mingles too many curses and execrations with her arguments." By contrast, "Miss More's performance breathes a more philosophic spirit, and appears in a more elegant garb." Badcock's judgment may have been clouded by his class prejudice, but the reviewer in the *Critical Review,* which was not generally supportive of abolition, did their best to rise above their preconceptions. "Religion, philosophy, and poetry" have united "to interest the reason, feelings, and fancy" in favor of "the unhappy negroes." More was "no despicable advocate" for their cause, the reviewer thought, and her "sentiments are no less just than happily expressed." Despite such faint praise from contemporaries, Basker contends that More "may have done as much as any writer to spread antislavery ideas among certain classes of the British reading public," while Wood notes that "More's abolition verse was massively popular and is the most fluent and intellectually the most hidebound verse to come out of mainstream state-backed Evangelical abolition." More "*must* now be read precisely because her assumptions regarding race, slavery, religion, and the economics of Empire are so far removed from the catechisms of contemporary Western liberalism." As Wood suggests, the poem is important precisely because it is such a central part of the "authorized" abolitionist literary canon.[19]

William Cowper had already commented on slavery in his celebrated poems *Charity* (1782) and *The Task* (1785). In his history of the abolition movement, Clarkson implies that Cowper volunteered "The Negro's Complaint" but, despite his previous verses on slavery, it appears Cowper had in fact resolved not to take part in the abolition campaign. There were, however, behind-the-scenes machinations that Clarkson was probably not aware of. According to a letter dated 21 March 1788 from Cowper to his cousin Lady Harriett Hesketh (1733–1807), he wrote the poem at the request of Lady Jane Balgonie (1757–1818), the daughter of evangelical Clapham sect philanthropist and abolitionist John Thornton (1720–90), one of the nation's wealthiest men and a subscriber to Cowper's ongoing translation of Homer. Lady Balgonie addressed the letter not to Cowper but to his friend and neighbor John Newton (1705–1807), the slave-trader-turned-abolitionist whose church career Thornton had supported. Lady Balgonie told Newton "We had some Gentlemen employed about the abolition of the Slave-trade with us the other day, they are very desirous of some good ballads to be sung about the streets on that subject, which they mean to print and distribute, and think they might be of use to the cause. If you think Mr. Cowper could by your means be prevailed on to do this for them, they would be extremely obliged to him."

Newton passed the note to Cowper, who reproduced it in his 21 March letter to Lady Hesketh. The levers of patronage thus pulled, Cowper set to work, albeit with some skepticism about the power of balladry to effect political change: "I do not perfectly discern," he wrote, "the probable utility of what I have done, for it seems an affair in which the good pleasure of King Mob is not likely to be much consulted; but at least it can do no harm." Cowper's doubts notwithstanding, the poem was published in the London newspaper *Stuart's Star, and Evening Advertiser* on 2 April 1789. Despite his initial reluctance, "The Negro's Complaint" became not only Cowper's best-known antislavery poem but perhaps the most famous of all poetic contributions to the abolition campaign, reprinted numerous times across the eighteenth and nineteenth centuries on both sides of the Atlantic.[20]

Although Roscoe's, More's, and Cowper's poems all emerged from the same personal and political networks, which had the abolition committee at their core, they present diverse arguments against the slave trade and speak in different poetic forms. Cowper offers a ballad intended for a popular audience, Roscoe emulates the Christian epic of Milton's *Paradise Lost,* while More is writing somewhat in the philosophical spirit of Alexander Pope's *Essay on Man.* Both Roscoe and More write in the voice of an omniscient third-person narrator, while Cowper ventriloquizes the voice of an enslaved African on a sugar plantation. Cowper and More consider both the slave trade and plantation slavery, while Roscoe is solely interested in exposing the slave trade. Roscoe and Cowper elide their sources and inspirations, while More offers a small selection of footnotes to support her argument. What they have in common is that they deploy the central tenets and key tropes of the claim that slavery is unnatural. These start with simple assertions but move on to more theorized arguments about human nature or natural or divine justice, scenarios in which God and/or nature are represented as actively intervening to expose or disrupt the slave trade, depictions of damaged environments, and the contention that plentiful African natural resources are being misappropriated.

Both More and Roscoe explicitly assert the idea that the slave trade is unnatural. More describes it as "Nature confounded" and an "unnatural deed." The word *unnatural* appears three times in the first book of *Wrongs of Africa,* most memorably when, emulating John Milton, Roscoe asks:

> But say, whence first th'unnatural trade arose,
> And what the strong inducement, that could tempt
> Such dread perversion?

The inducement, he goes on to show at length, is avarice, fomented by Satan himself, but the statement that the slave trade is unnatural is explicit, as it is some pages later when he argues that white slave traders in Africa incite "contending armies, unprovok'd / By previous wrong, to wage unnatural war." The proslavery claim that the Africans bought by European slave traders were legitimate captives of legal wars was familiar, as was the abolitionist counterclaim that these wars were got up by slave traders to feed their market, but Roscoe elevates abolitionist arguments that these wars were illegal to the assertion that they were unnatural. He compares "the white deceiver" who "had sown / The seeds of discord" with the devil of *Paradise Lost:*

> Artful, and fair, and eloquent of speech,
> Was the first tempter, that in Eden's groves,
> Guiltless before, brought sin, and pain, and death:
> And fair, and artful, were the cultur'd train,
> That wound the snare round Afric's thoughtless sons,
> And dragg'd them to perdition.

Deception is both satanic and unnatural, and this theme unites "Part the First" of the poem. Playing on the old adage that one cannot serve both God and Mammon, and adjusting for the age of sensibility in which he was writing, Roscoe constructs a personified figure of Avarice who is somewhat related to the Miltonic figure of Mammon, under whose influence men "Ransack'd the Center, and with impious hands / Rifl'd the bowels of thir mother Earth." Avarice, in Roscoe's view "the foulest fiend that ever stalk'd / Across the confines of this suffering world," is easy enough for a poet to condemn, but Roscoe is more concerned when duplicitous Avarice is found to "personate the form divine / of soft compassion." Like the Satan of *Paradise Lost,* the "Artificer of fraud" who was "the first / That practisd falshood under saintly shew," slave traders are not, as they claim, saving African prisoners of war from a worse fate, nor are they demonstrating compassion when they care for captives in the Middle Passage with decent food and medical attention. This "sympathy," argues Roscoe, lasts only as long as the voyage and is in any case fraudulent: "The hated power / Of unrelenting avarice" had "late claim'd unnatural union" with sympathy. If hard-heartedness is unnatural, so much more so deception. In Roscoe's analysis, the slave trade is unnatural since it is founded on the "dread perversion" of natural sensibility by commercial avarice.[21]

Both More and Roscoe also attempt more highly theorized justifications for the argument that slavery is unnatural based on current scientific thinking as well as abolitionist polemic. More explicitly cites Ramsay and implicitly refers to Benezet and Clarkson, but the poem's guiding metaphor, with which it opens and closes, compares the abstract idea of liberty with the sun. Unlike the actual sun, however, the "penetrating essence" of liberty is not constrained by geophysical phenomena such as gravity, seasons, or the curvature of the Earth. "Since there is no convexity in MIND," More reasons, it seems wrong that the "bright ray" of liberty "shou'd'st ne'er irradiate all the earth." She asks, "While Britain basks in thy full blaze of light, / Why lies sad Afric quench'd in total night?" The answer she offers is that Africa's supposed lack of liberty is in fact caused by British slave traders. She develops this observation later in the poem, but it is notable that she chooses a global metaphor contrasting the twin pillars of Enlightenment thought—Newtonian physics and personal freedom—to represent a global problem. The preface to *Wrongs of Africa,* written by James Currie (1756–1805), also aligns abolitionism with the illumination of scientific progress. "Discoveries in science" argues Currie, "are very rapidly increasing the power, amending the condition, and enlarging the views of mankind; and the close of the eighteenth, like that of the fifteenth century, will probably be marked in future times, as a period in which a sudden accession of light burst upon the human mind." Currie's prediction that scientific progress would inevitably cause "hatred and bloodshed [to] be exchanged for confidence and peace" may have sounded as naïve in 1787 as it does today, but his preface goes on both to invoke the suffering of African people taken captive in the slave trade and to offer, as any good piece of scientific writing would, statistics supporting his case. "That representations such as these should have no influence in a country where men have heads to reason and hearts to feel," he argues, "is impossible." More continues her appeal to natural philosophy, or science, in lines that appear to constitute a pointed rebuff of racial theorists such as Edward Long. The couplet "Perish th'illiberal thought which wou'd debase / The native genius of the sable race!" is balanced by another exclaiming, "Perish the proud philosophy, which sought / To rob them of the pow'rs of equal thought!" The couplets are followed by a series of rhetorical questions in which the poet probes her readers' understanding of human nature, concluding with a forceful assertion of the spiritual and emotional equality of Europeans and Africans, substantiated by a long footnote from the section of James Ramsay's *Essay* in which Ramsay had refuted the ideas of Long and others. In the case of both Roscoe and More, aligning their poems with progress in natural history and natural philosophy as well as with abolitionist polemic allows the poets to argue that opposition to the slave trade is not merely a political position but an inevitable necessity since the trade is a "dread perversion" of nature that must be set right in the natural course of things.[22]

While both More and Roscoe are keen to align their abolitionist principles with current scientific thinking, both are also Christians who view the laws of nature as synonymous with the laws of God. More's *Slavery* ends with a vision of "the cherub Mercy" joining with the spirit Liberty to conquer the giant Oppression such that "FAITH and FREEDOM spring from Mercy's hands." Roscoe's *Wrongs of Africa* ends with an extended apostrophe to the "divine nymph" Freedom, who conquers personified Cruelty, Avarice, and Sophistry, culminating in a prayer for "truth, and justice, and unbounded love" that ascends "to the mercy-seat of God."[23]

William Cowper, by contrast, steers away from these labored allegories to offer a more succinct although no less metaphorical account of the injustice of slavery. Cowper echoes More's assertion of the spiritual and emotional equality of Africans and Europeans with the "Negro" of the poem's title arguing, "Fleecy locks and black complexion / Cannot forfeit nature's claim." Whereas More's rhetorical questions are answered by the poet herself, in Cowper's poem, God answers through the medium of extreme weather:

> Hark! He answers!—Wild tornadoes
> Strewing yonder sea with wrecks,
> Wasting towns, plantations, meadows,
> Are the voice with which he speaks.
> He, foreseeing what vexations
> Afric's sons should undergo,
> Fixed their tyrants' habitations
> Where his whirlwinds answer—"No."

One of the most traditional ways in which authors could signal divine displeasure is through environmental catastrophe, including crop failure, plague, earthquake, and flood. In the spirit of this tradition, Cowper conjures up some appalling weather to destroy ships and towns and lay waste to plantations and meadows. In this poem, however, God's wrath is not merely a temporary response to a single transgression. In Cowper's conception it is not passing weather events but the perilous Caribbean climate itself that signals that God has permanently rendered the region unsuitable for colonization—or indeed, any sort of development. The whirlwinds are simply a dramatic and recurring reminder of God's fixed intention. Although the stanza reads initially as a rather egregious instance of the pathetic fallacy, it links the poem to the tradition of natural theology in which interpreting the natural world could offer the theologian insights into the nature and intentions of God. Setting enslaved African laborers to work in the Caribbean, implies Cowper, is both ungodly and unnatural. Roscoe clearly thought the same of the slave trade, depicting the unfeeling captain of a slave ship who "seest unmov'd / The lightning's glare, and hear'st the thunders roll" when crossing "th'insulted ocean." Despite the natural obstacles placed in his way to protect Africa from Europe, "in vain the elemental fury rag'd" and the captain reaches his destination, clearly against the wishes of nature and its creator.[24]

Cowper's whirlwinds dramatize the damage that the environment could wreak on an inappropriately sited and organized colony, but environmental despoliation was not God's prerogative alone. Roscoe depicts African society as "cheerful" and naturally productive before "the wasteful rage / Of European avarice chang'd the scene." Roscoe was not alone among eighteenth-century observers in recognizing that humans could damage environments, either by poor management, deliberate destruction of resources, or actions that caused depopulation. While modern environmentalists would be more apt to identify overpopulation as a threat, to eighteenth-century observers who believed, in crude terms, that the earth was a garden in need of human cultivation, depopulation could lead to catastrophic impacts on land management. This was especially pertinent in the debate over the slave trade, which depopulated African coastal regions. But mismanagement could take place on both sides of the Atlantic. The "Negro" of Cowper's poem asks, "Why did all-creating nature / Make the plant for which we toil?" Although contemporary readers may have understood this to mean that the African narrator is simply ignorant of God's plan, another implication is not that God made a mistake in creating sugarcane but rather that European settlers made a mistake in introducing it, and the enslaved workers who cultivated it, into the Caribbean. The plant itself becomes the source of human misery: "Sighs must fan it, tears must water, / Sweat of ours must dress the soil." The sighs, tears, and sweat are metonyms for the suffering slavery causes, but the soil itself is transformed. In Cowper's conception, the unnaturally introduced crop becomes the locus of both human and environmental damage.[25]

Images of despoliation in the poems are often contrasted with descriptions of the plenitude of African nature before the coming of the Europeans. *Wrongs of Africa* in particular offers a vision of a bountiful nature corrupted by avarice in which the greedy slave trader, "gratified beyond his utmost wish, / Debars another from the bounteous store."

From her exhaustless springs the fruitful earth
The wants of all supplies: her children we,
From her full veins the grateful juices draw,
With life and health replete; nor hard return
She at our hands requires, nor more than suits
The ends of health and pleasure; yet bestows
On all her offspring with a parent's love
Her gifts impartial.

This prelapsarian image represents the African natural environment as a Garden of Eden brimming with resources for human consumption. It recalls the extended description of Eden in book 4 of *Paradise Lost,* in which Milton had not only depicted the bounty and beauty of divine providence but had also contrasted Satan's entry into Eden with mariners who sail "Beyond the *Cape of Hope,* and now are past / *Mozambic.*" Roscoe picks up Milton's pun on being beyond hope when on the coast of Africa in "Part the Second" of the poem, where he likens the descent into the "crouded holds, and loathsome caverns" of a slave ship to a descent into a Miltonic hell, where "dread despair, / And anguish inexpressible" signal that "hope's slender thread was broke." This is consistent with his repeated emulation of *Paradise Lost.* Earlier, as we have seen, he had paraphrased Milton when he asked, "But say, whence first th'unnatural trade arose?" and pondered what could be "the strong inducement, that could tempt / Such dread perversion?" Satan, it is repeatedly implied, is at the root, but the human cause is the slave trade, driven by avarice, which has disrupted the harmonious relationship with nature that Africans, "cheerful natives," enjoyed before "European avarice chang'd the scene."

—Strangers alike to luxury and toil,
They, with assiduous labour, never woo'd
A coy and stubborn soil, that gave its fruits
Reluctant; but on some devoted day,
Performed the task, that for their future lives
Suffic'd, and to the moist and vigorous earth
The youthful shoots committed: fervid suns,
And plenteous showers, the rising juices sent
Thro' all the turgid branches; and ere long,
Screen'd from the scorching beam, beneath the shade
Himself had rais'd, the careless planter sat;
And from the bending branches cropt the fruit;

More grateful to his unperverted taste,
Than all that glads the glutton's pamper'd meal.

Roscoe again alludes to *Paradise Lost* in the final line; on his first visit to Eden, Satan takes the form of a cormorant (*Phalacrocorax carbo*), a traditional symbol for gluttony. Roscoe unexpectedly describes the African farmer as a "planter," which draws a sharp contrast between the unfeigned simplicity of the African farmer, working in harmony with "the moist and vigorous earth," and the unnatural planter of the Caribbean, forcing enslaved people to tend a reluctant crop, all to feed the perverted taste of a glutton. The comparison is another, albeit elliptical, implication that slave traders and slaveholders are satanic. Milton is not the only referent, however. The image of Africa as prelapsarian was widespread in the travelers' literature that Anthony Benezet had consulted and, like Day before him, Roscoe seems to be alluding to the image of an Edenic Africa that Benezet had extracted from the works of Michel Adanson. The references are not entirely external, however. Throughout "Part the First," Roscoe speaks repeatedly of the perversion of nature. The "detested joys" of the slave trader are "some perversion of each nobler sense / Indulgent nature gave thee," the slave trade is a "foul plague" from Europe "Perverting good, to evil: at the sight / Nature recoil'd," while the "unnatural trade" is a "dread perversion." Against these are contrasted the "unperverted taste" of the "careless planter" on his farm in Africa. The poem's consistent message is that the slave trade is as damaging to nature as it is to liberty.[26]

This is More's message too. Toward the end of *Slavery,* she returns to the language of science, albeit with the personification of mercy as a sentimental "cherub," but she develops the idea of liberty via an image of the spirit of mercy rejuvenating an African landscape blighted by European slave traders:

As the mild Spirit hovers o'er the coast,
A fresher hue the wither'd landscapes boast;
Her healing smiles the ruin'd scenes repair,
And blasted Nature wears a joyous air.
She spreads her blest commission from above,
Stamp'd with the sacred characters of love;
She tears the banner stain'd with blood and tears,
And, LIBERTY! thy shining standard rears!

The passage dramatizes the refreshing impact that liberty would have on a withered, ruined, and blasted landscape. Her portrayal of Africa draws on and inflates the description of the environmental impacts of the slave trade found in Benezet's *Some Historical Account,* although the hyperbole is necessary to achieve the poetic intensity that a polemical poem requires. Her argument that the slave trade is damaging the African environment refers back to her suggestion ten pages earlier that gold is "better gain'd, by what their ripening sky, / Their fertile fields, their arts * and mines supply." The mid-sentence asterisk leads to a footnote asserting, "Besides many valuable productions of the soil, cloths and carpets of exquisite manufacture are brought from the coast of Guinea." The footnote suggests that More had looked into Thomas Clarkson's traveling chest of African produce. This impression is reinforced three pages later in another footnote, this time to the line "When the sharp iron * wounds his inmost soul," which is footnoted with the statement "This is not said figuratively. The writer of these lines has seen a complete set of chains, fitted to every separate limb of these unhappy, innocent men; together with instruments for wrenching open the jaws, contrived with such ingenious cruelty as would shock the humanity of an inquisitor." The two footnotes imply that More had been to one of Clarkson's events during his visit to Bristol in the autumn of 1787 when, in the words of More's biographer, the two "had some contact," even if neither kept a detailed record of the encounter. Equally, the poetic representation of an Africa teeming with misappropriated natural resources may have come via Clarkson's writing or Benezet's, and it echoes Roscoe's depiction of an Africa that is not only Edenic but also prodigal in offering up the produce of its "moist and vigorous earth."[27]

All three poems, in their various ways, offer a vision of an unnatural trade servicing an unnatural plantation regime. The extended narratives of Roscoe's and More's poems allow them to develop complex arguments and lengthy metaphors. Both either implicitly or explicitly refer to the arguments of Benezet and Clarkson, and both are united not only in asserting that the slave trade is unnatural but in demonstrating the many forms this "dread perversion" takes. More and Roscoe clearly had access to the antislavery essays by Benezet, Ramsay, and Clarkson, but Cowper's engagement with abolitionist literature in "The Negro's Complaint" is neither specific nor referenced, and the poem has all the hallmarks of a spontaneous composition. Nevertheless, in this most famous of all abolitionist poems, Cowper represents the slave trade not merely as unnatural but also as directly in contention with nature. The poem's conclusion, a demand from the poem's enslaved narrator to the British reading public to "prove that you have human feelings, / Ere you proudly question ours!" chimed with the abolitionist logo of an enslaved

African in chains on his knees asking the implied white observer, "Am I not a man and a brother?" The poem's rallying cry may have accounted for much of its popularity but, while its British readers are asked to prove in the here and now that they have human feelings, the poem's broader implication is that the slave trade is so out of kilter with the natural world that the forces of nature will ultimately step in to destroy it. This view was very much in accord with abolitionist thinking in the 1780s. "Slavery," argued Ramsay in 1784, "is an unnatural state of oppression on the one side, and of suffering on the other; and needs only to be laid open or exposed in its native colours, to command the abhorrence and opposition of every man of feeling and sentiment." Although it would turn out to be unduly optimistic, at least in the medium term, Ramsay's opinion that slavery was unnatural and would naturally be abolished as soon as it was in the public eye was shared by most abolitionists in the 1780s. It is powerfully reflected in the antislavery poetry of the decade.[28]

Conclusion

"AN INEXHAUSTIBLE MINE OF WEALTH"

Despite all the arguments that had been made against it, including that it was unnatural, the British slave trade was not abolished in the early 1790s. Proslavery parliamentarians employed delaying tactics, public attention was diverted to the international situation following the French Revolution, and abolitionists themselves began to weary of the fight. More than a decade passed with little progress, but in 1806 a window of political opportunity was opened and the arguments that had previously been made so cogently against the slave trade were presented again to the reading public, sometimes by veterans of the campaign and sometimes by members of a new generation. Many felt that there was little more to be said on a matter that had been decided, if not carried through, fifteen years earlier. Introducing the crucial second reading of the Slave Trade Abolition Bill to the House of Commons on 20 February 1807, at which the bill was substantially debated, Charles Grey (1764–1845), at that point Lord Howick though still sitting in the House of Commons, "said, he never came to the discussion of any subject with more embarrassment, than on the present occasion. The question had been so often agitated, that every hon. member could not but be acquainted with all its details." William Wilberforce agreed that he had not heard from the witness called on behalf of planters and slave traders "a single new point, except that which had been justly termed by the noble lord matter of opinion. Evidence was therefore unnecessary."[1]

As several members pointed out, the debate in Parliament in 1807 revolved primarily around "justice, humanity, and sound policy" and there was relatively little discussion of whether the trade was natural or unnatural. The same was not true outside of the House, where pamphleteers and poets returned to the rhetoric of the previous decades. Thus the veteran abolitionist William Smith (1756–1835), in *A Letter to William Wilberforce, Esq., MP, on the Proposed Abolition of the Slave Trade,* called the slave trade a "natural" but "disgusting . . . union of Moloch and Mammon," arguing that no human could ever be a slave because "as long as he remains a rational, moral, accountable creature, it arises out of the essence of his nature that he cannot be the proper subject of barter and sale." Smith added that it was "impious" for slave traders and slaveholders "to erect municipal institutions in opposition to the eternal law of nature." The anonymous *Gleanings in Africa,* a collection of letters on slavery and natural history in southern Africa published in 1805, makes the point repeatedly. "The vast theatre of the moral world" does not "exhibit a more disgusting spectacle than the existence of slavery, that odious and detestable system, which gives to one man so undue and unnatural a power over another." Slavery "is the degradation of humanity and a direct violation of the laws of Nature." Poets also continued the theme. *The Slave,* written in 1807 by the churchman Richard Mant (1776–1848), serves up every convention of abolitionist poetry as a fascinating illustration of the way in which convention becomes cliché. Mant depicts a lush and paradisical Africa where "peace and plenty dwell, for man is free." Slave traders despoil this Eden such that "no joyous hail / From grateful Afric greets fair Albion's sail." There are "groans instead, and curses deep resound, / And death and desolation gloom around." These slave traders capture an innocent African farmer and "rend with impious force apart / The ties, which nature wind around his heart" before God shows his displeasure "when from the south the mad tornadoes rise." Arriving in Jamaica, the captives may dream of another tropical paradise where "The humming bird, to charm thy wond'ring eyes, / Bright to the sun shall show his rainbow dies." Instead, there is a vision of paradise lost, or rather paradise taken away, as the poet offers, " '* From morn till noon, from noon till dewy eve,' / No comfort cheer thee, and no rest relieve." The asterisk calls out a footnote stating, "A line from Milton" in case the reader fails to spot his mangled misquotation. The poem is not good, but it serves to illustrate how deeply ingrained the images of prodigious but despoiled tropical landscapes had become in abolitionist verse by the early years of the nineteenth century and suggests that, whatever legal principles engaged parliamentarians, abolitionist popular culture remained wedded to the idea of slavery as a "dread perversion of nature."[2]

Such ideas persisted after 1807, and numerous examples can be found from the debates about the abolition of British colonial slavery in the 1830s, those about American slavery in the 1850s and '60s, and even in discussion of modern slavery. Slavery may have been outlawed across the British Empire, but it was frequently replaced by other forms of coercive labor. Chattel slavery remained widespread in North and South America into the late nineteenth century—it remained legal in the United States until 1863, in Cuba until 1886, and in Brazil until 1888. Elsewhere, slavery in its various forms remained legal in some countries until the late twentieth century. It was not outlawed in Mauritania until 1981, even though the Universal Declaration of Human Rights had proclaimed in 1948, "No one shall be held in slavery or servitude; slavery and the slave trade shall be prohibited in all their forms." Today, although illegal throughout the world, slavery remains widespread and common. Writers on modern slavery do not often use the word *unnatural,* with its connotation of a fixed understanding of what nature permits, but they frequently note that the attitudes that perpetuate modern slavery are correlated with those that ignore impacts on climate, wildlife, or habitats, while exploitation of people often goes hand in hand with exploitation of the environment. In modern Brazil, notes Kevin Bales, slavery and environmental destruction sweep like a tidal wave across the Amazon. Here, "as the native ecosystem and peoples are uprooted, displaced workers, even the urban unemployed, become vulnerable to enslavement." The people here "are completely under the control of their masters. The wave is carrying slavery with it. The land ahead is still exploitable, the land behind is stripped, and when all the land is stripped the slaves will be discarded." While the language has changed, Bales is representing slavery as a "dread perversion of nature" no less than were eighteenth-century abolitionist authors.[3]

The argument that the slave trade was unnatural had longevity, then, but it remains legitimate to ask if it was successful and if it had further impacts. Clearly, the arguments against first the slave trade and then slavery itself were successful, albeit more slowly than the activists of the 1780s might reasonably have hoped. British public opinion appears to have firmly turned against both by 1789, when the slave trade was first debated in Parliament, and it was only dogged parliamentary resistance by the West India Interest in 1792 and beyond that blocked immediate anti–slave trade legislation. The lack of progress in the quarter century between the end of the British slave trade in 1807 and the end of slavery in the British Empire in the 1830s can also, at least in part, be attributed to the political process, both within and without the abolition campaign, rather than to wider public opinion. The decision by abolitionists such as Wilberforce and James Stephen (1758–1832) to focus on slave registration in the years after 1807, combined with the lack of a formal antislavery society before 1823, meant that opportunities to mobilize popular antislavery sentiment were lost. When opposition to slavery again became the focus of sustained political campaigning in the 1820s, the arguments again focused on its cruelty and inhumanity, its incompatibility with both Christian principles and British notions of liberty, and its economic and strategic failures. The idea that slavery was unnatural, as we have seen, continued to be stated both explicitly and implicitly and remained a valuable part of the discourse of antislavery—but precisely how valuable is impossible to quantify without performing the impossible experiment of rerunning history with a different set of discourses.

The eighteenth-century notion of the slave trade as unnatural did, however, have a somewhat surprising consequence. Abolitionists of this period consolidated and popularized the growing opposition to the economic and environmental models that the plantocracy advocated, but at the same time they brought discussion of tropical environments into a mainstream public arena. This was a mixed blessing. On the one hand, they advanced antislavery sentiment in tandem with awareness of new habitats and organisms. On the other, they familiarized readers with the productive potential of exotic landscapes and by so doing reinforced the idea that tropical lands could, and should, be subject to British colonial management, or at least be open to trade on terms that were advantageous to the British. An unexpected outcome, therefore, of a century of yoking together discussion of slavery and nature was that while slavery and the slave trade could be viewed as unnatural, exploiting the natural resources of Africa and other tropical regions could be seen as a natural remedy. From the 1780s onward, therefore, abolitionists helped mobilize public support for imperialism, sometimes inadvertently, but all too often with great enthusiasm. The natural history model adopted by abolitionists who wrote about Africa and the Americas offered a real, scientifically grounded, and seemingly virtuous justification for exploration, appropriation, and colonization, since, they could argue, extending the bounds of knowledge could only be a good thing, especially when conducted in tandem with humanitarian antislavery activism. In reality, the type of knowledge most sought after was the sort that would empower and enrich the nation. As a result, abolitionists were paradoxically among the first to promote the colonization of Africa explicitly to exploit Africa's natural resources.

Many advocated replacing the slave trade with free trade. Allowing African societies to freely trade their produce and commodities would, so the argument went, obviate the need for them to sell their own people to acquire capital to buy foreign goods—while at the same time boosting exports of British manufactures. The economic theory behind this thinking was largely derived from Adam Smith's (1723–90) influential *Wealth of Nations,* published in 1776, in which he also argued that "the work done by freemen comes cheaper in the end than that performed by slaves." Smith's stance on slavery no doubt facilitated the adoption of his views on free trade by abolitionists. The issue was that, to promote trade in African products and commodities, one would need to know what they were. For information such as this, one could always turn to natural history. This is precisely the approach Anthony Benezet takes in *Some Historical Account of Guinea* where he declares his intention to "give some account of the several divisions of those parts of Africa from whence the Negroes are brought, with a summary of their produce." While many of the authors he cites are not career naturalists, they are educated travelers whose observations are grounded in the prevailing methods of natural history. Others, like Michel Adanson, were indeed important naturalists. Very quickly, their combined evidence builds up a picture of Africa as a land where "the country is rich, and the commerce advantageous, and might be greatly augmented by such as would cultivate the friendship of the natives." Benezet's *Some Historical Account* ends with an appeal to the commercial instincts of British and American readers that was widely imitated:

> Africa has about ten thousand miles of sea coast, and extends in depth near three thousand miles from east to west, and as much from north to south, stored with vast treasures of materials, necessary for the trade and manufactures of Great-Britain; and from its climate, and the fruitfulness of its soil, capable, under proper management, of producing in the greatest plenty, most of the commodities which are imported into Europe from those parts of America subject to the English government; and as, in return, they would take our manufactures, the advantages of this trade would soon become so great, that it is evident this subject merits the regard and attention of the government.

On the one hand this can be read as a progressive vision of free trade, international commerce, and opportunities for African development. On the other hand, we can read it as a call to exploit the natural resources of a continent by colonization, since it is implied that only Europeans could offer the "proper management" necessary to work this economic miracle. Natural history, humanitarianism, and economic optimism came together to create the formula that would be used to justify European colonialism in Africa.[4]

Benezet's argument was reproduced in different forms throughout abolitionist literature. "Nothing can be more clearly shewn," argued Thomas Clarkson in his *Essay,* "than that an inexhaustible mine of wealth is neglected in Africa, for the prosecution of this impious traffick; that, if proper measures were taken, the revenue of this country might be greatly improved, its naval strength increased, its colonies in a more flourishing situation, the planters richer, and a trade, which is now a scene of blood and desolation, converted into one, which might be prosecuted with advantage and honour." This belief animated Clarkson throughout his career, such that from the following year onward, he began to bring his traveling chest of African natural produce to public-speaking engagements to demonstrate the riches that the continent might yield. Clarkson was not alone. At the start of *The Interesting Narrative,* Olaudah Equiano offers a representation of Africa as "uncommonly rich and fruitful." At the end, in words that echo Clarkson's conception of Africa as "an inexhaustible mine of wealth," he asserts, "A commercial intercourse with Africa opens an inexhaustible source of wealth to the manufacturing interests of Great Britain. . . . Population, the bowels and surface of Africa, abound in valuable and useful returns; the hidden treasures of centuries will be brought to light and into circulation. Industry, enterprize, and mining, will have their full scope, proportionably as they civilize." As we saw in the introduction to this book, Wilberforce took this argument to Parliament in May 1789. "The continent of Africa," he argued, "furnishes several valuable articles of commerce highly important to the trade and manufactures of this Kingdom, and which are in great measure peculiar to that quarter of the globe; and . . . the soil and climate have been found, by experience, well adapted to the production of other articles." Abolishing the slave trade and replacing it with a trade in Africa's natural resources based on "true commercial principles" would not only "make reparation to Africa, so far as we can" but might reasonably be expected to promote "the progress of civilization and improvement on that continent." Abolitionists who promoted the idea of Africa as a continent rich in natural resources inadvertently fueled the sense that riches were to be had there. Such attitudes, while they were in the short term useful in promoting the anti–slave trade message, were in the long run closely allied to ideologies and national interests that viewed supposedly underdeveloped Africa as a natural storehouse ripe for colonization.[5]

Abolitionists were deeply immersed in natural history. In turn, natural historians were influenced by abolitionism. This is especially evident in accounts of exotic floras and faunas as British explorers and colonists spread throughout America, Asia, and the Pacific. Almost all the authors of these texts would have had at least some exposure to both antislavery and proslavery ideas, and many incorporated antislavery ideas into their accounts of wildlife wherever they observed it. Some of these have recently come under the scrutiny of historians. Henry Smeathman (1742–86), for example, who studied West African insects and was later a leading figure in the scheme to resettle London's black poor in Sierra Leone, had combined descriptions of African slavery with accounts of the African termite. His works and colorful life have attracted attention in recent years, although his natural history, despite its attention to slavery, is not directly linked to the emerging abolition campaign. John Gabriel Stedman (1744–97), whose *Narrative of a Five Years Expedition against the Revolted Negroes of Surinam* was published in London in 1796, was professionally a soldier but devoted much of his time, and much of his *Narrative,* to the flora and fauna of Surinam. The book was both influenced by the debates over the slave trade and contributed to further those debates. Some celebrated naturalists attempted to sit on the fence. Joseph Banks (1743–1820) was an enthusiastic advocate for British colonialism who actively supported the African Association which, from the 1780s, promoted exploration and discovery in Africa. This association was to some extent welcomed by abolitionists, who felt that greater knowledge of Africa and its people, coupled with free trade, would tend to diminish the importance of the slave trade. "To begin with," notes John Gascoigne, "Banks saw the aims of the Association as the fostering of exploration. However, as time went on and the conflict with the French became more acute, he became more inclined to link the Association's activities with more overtly imperialistic aims." Earlier, in 1787, Banks had been instrumental in persuading the government to send out an expedition to the Pacific Ocean to gather breadfruit (*Artocarpus altilis*), at the time an almost legendary tree, which, it was thought, could be cultivated in the West Indies as a cheap and reliable food for the enslaved population. The voyage did not end well. Banks's choice of William Bligh to command HMS *Bounty* led to one of the most famous maritime incidents of the eighteenth century. The mutiny notwithstanding, the voyage was a clear attempt by one of the most celebrated naturalists of the age to find a way to sustain rather than undermine plantation slavery. A second voyage was more successful and, by 1793, breadfruit had been introduced to the Caribbean, where it remains on menus to this day. Banks's private correspondence suggests that, while he hoped for the long-term decline of slavery, he did not support direct attempts to hasten that decline. Overall, according to Gascoigne, this "was an issue about which Banks was rather ambivalent, oscillating between his Enlightenment (and perhaps, too, residual Christian) faith in the ultimate equality of humankind and the imperatives of Empire." The breadfruit episode remains, however, one of the clearest efforts to unnaturally sustain an unnatural empire, in which a natural historian leads an expedition to introduce an exotic crop into a colonized island to support an enslaved population that labors to grow another exotic introduced crop for the benefit of a metropolitan market.[6]

Much abolitionist writing relied heavily on the work of Anthony Benezet. As Eric Herschthal has pointed out, however, "one of the problems with Benezet's *Historical Account* was that the natural histories it cited were decades old." This in fact probably made little difference to many supporters of the early abolition movement, who were not themselves scientists and for whom the moral case that Benezet rested on the testimony of travelers and naturalists would have outweighed the issue of the freshness of the evidence. Benezet's arguments, his rhetorical style, and the fact that he had a network of antislavery Quakers behind him all counted for more in the political marketplace. Herschthal implies, but does not directly establish, that dissatisfaction with the science in *Some Historical Account* was one of the motivations behind English Quaker John Fothergill's (1712–80) decision in 1771 to support Smeathman's expedition to Sierra Leone, which Banks also funded. In fact, if Fothergill was indeed reacting against Benezet, it is more likely that one of Benezet's earlier publications was the spur, almost certainly *A Caution and Warning to Great Britain and Her Colonies,* published in Philadelphia in 1766 and reprinted the following year in London. Nevertheless, the expedition both directly and indirectly led to the Sierra Leone settlement project in which many abolitionists, including Granville Sharp and Equiano, were directly involved.[7]

As Herschthal shows, numerous further expeditions to Africa after the 1770s involved those who held antislavery views. While Benezet, Clarkson, and many other abolitionists facilitated the idea that Africa was rich in exploitable resources, some eighteenth-century abolitionists directly called for widespread European colonization in Africa. A notable example was Carl

Bernhard Wadström (1746–99), a disciple of the Swedish mystic Emanuel Swedenborg (1688–1772), who in 1787–88 participated in a surveying voyage to Africa on behalf of the Swedish government. Wadström aimed to establish a supposedly enlightened colony in West Africa to trade with Africans on equal terms. On the voyage, he witnessed firsthand the brutality of the slave trade and on his return to Europe became a celebrated abolitionist in Great Britain, giving evidence before Parliament on the strength of his *Observations on the Slave Trade,* published in London in 1789. In the opening three sections, Wadström offers a compelling abolitionist pamphlet containing firsthand accounts of the war and pillage that the slave trade caused in Africa. The fourth section is a natural history of the coast of Africa, covering climate, soils, animals, plants, and minerals, while the fifth considers the natural "impediments which will oppose the European Settlements on the Coast of Guinea." Many of Wadström's claims about the soil and geology of Africa are generalizations made on the basis of very limited observations taken at the coast, while his method as a natural historian is generally anecdotal and imprecise, with an emphasis on those plants and animals that are likely to be useful to Europeans and a lack of any systematic biological knowledge. Nevertheless, the natural history passages offer a revealing insight into the mind of "colonizing abolitionist" at the end of the eighteenth century. At the start of his book, he sets out his intention to write "of the manner in which the negroes are treated by the Europeans; but more particularly of the possibility of improving, by cultivation, the fruitful soil of Africa." Thus, in the natural history chapters of the book he focuses almost exclusively on resources that might be of use to Europe and finally advocates an end to the slave trade by growing cash crops in Africa instead of America. Like Benezet and Clarkson, he emphasizes the prodigality of the African environment, noting that "sugar-canes grow wild in many places, which with a little cultivation might be rendered extremely valuable and productive. The same may be said of the tobacco-plant. Several species of cotton are also spontaneously produced by this excellent soil." All three plants he mentions are economically important, but while Levant cotton (*Gossypium herbaceum*) may be found in semi-arid regions of West Africa, the other two are introductions. A fourth plant, indigo (*Indigofera tinctoria*), is introduced as an extended metaphor for African people taken into slavery and transplanted across the Atlantic: "Indigo of different kinds also grows wild, and in such quantities, as to be a very troublesome weed in the rice and millet fields. What a strange inversion of nature does not man, actuated by the most extravagant and most ridiculous selfishness, every where labour to effect? What necessity is there for exiling this plant from the soil and climate which nature has assigned it, in order to transplant it into a country, where it is far from thriving so well as in its native place, and where it fails every third or fourth year?" Since neither people nor indigo plants thrive after transplantation, they should be allowed to flourish in their native soil. There is some unintended irony since indigo is a cosmopolitan plant that may be native to Asia, not Africa, but Wadström echoes commonplace abolitionist sentiment when he calls this "a strange inversion of nature" that is stimulated by European avarice. Despite this, he does not actually advocate abandoning the luxuries of sugar, tobacco, cotton, and indigo. Rather, he explicitly calls for "a little cultivation" and in the remainder of the book he discusses the "impediments" to colonization that Europeans might face and the means by which they might be overcome. His conclusion problematically yokes together a call for settlement with strident abolitionist sentiment:

> Let us then form new settlements along the African Coast; settlements which shall have no other aim than that of inviting those nations to the riches which will arise from the cultivation of their own country, and thence the enjoyment of civilization, to both which they are capable of applying themselves with ardour and joy.—Let us thus on the wreck of tyranny raise altars to humanity. Let us give to this weak, timid, and ignorant people, a masculine and courageous education. Let us make them feel the nobility of their origin, that under our tuition they may become generous from sound political interest; and may they no longer be slaves, but men. Let us for our own part freely assist them in tilling the fine country they inhabit. Let us prove to those innumerable multitudes of men, by the force of example, that they possess the most fertile soil. Let us also, by example, teach them no longer to suffer themselves to be torn from their native shores. Let us teach them to shake off the irons, and to revenge themselves on the blind tyrants, who shackle them, by becoming more useful to them in a state of freedom.

This extraordinary passage explicitly advocates the European colonization of Africa, denigrates Africans as weak and ignorant, appears to hold them responsible for their own enslavement at the hand of Europeans, and charges Europeans with the burden of educating and civilizing Africa. It comes at the end of a book in which African produce and African soils have been the center of attention and which couched its colonial ambitions in the language of exploration and scientific investigation. It is not itself a work of natural history, but its colonial ambitions are underpinned by two centuries of work by colonial natural historians.[8]

Five years later, Wadström repeated his call for colonization of Africa in a lengthy and detailed book explicitly called *An Essay on Colonization,* much of which consists of a miscellany of quotations from authors, naturalists in particular, who discuss Africa. The book considerably expands both Wadström's testimony and his opinions. "The colonization of Africa," he argued, "is not only practicable, but, in a commercial view, highly prudent and adviseable." He continued to advocate strongly against the slave trade, but there was never any doubt whom he saw as the primary beneficiary of colonization: "The day, I hope, is not far distant when Africa will enrich Europe with the most lucrative commerce." His colonial model was paternalistic. He assumed that Africans were fundamentally as intellectually capable as Europeans, but that they were at an earlier stage of social and cultural development. "Societies may be divided into the *civilized* and the *uncivilized;* and the duties of the former to the latter are similar to those of parents to children," he argued, adding, "As the tutelage of children is a state of subjection; so it would seem that civilized nations have perhaps some right to exercise a similar dominion over the uncivilized, provided that this dominion be considered and exercised as a mild paternal yoke." He extended the metaphor further, suggesting that this dominion should cease once Africa "arrives at maturity" and is then at some pains to assure readers that "by this paternal dominion, I am far from intending any species of arbitrary power."[9]

Claims that paternalistic empire would be benevolent in nature and beneficial to Africans were repeatedly and increasingly made across the nineteenth century, but Wadström is unusual in laying at least some of the blame for the alleged inferior development of Africa at the door of European slave traders and, in particular, the impact they have on the African environment. "The slave-trade disturbs their agriculture," he argues, "for men will not be fond of planting who have not a moral certainty of reaping." By contrast, he shows that in areas "where there is no slave-trade, they have made great progress in agriculture." Such arguments depend in the final analysis of the relative abilities of European and African farmers and others to exploit the natural resources of the continent. While the first three chapters of the book dismiss "objections to colonization" before discussing the African character and the concept of civilization in general, after fewer than thirty pages the book turns to Africa's climate, soil, water, and produce. This is prodigious: in some parts, where the soil is good, "the vegetation is luxuriant to a degree unknown in the most fertile parts of Europe, and the trees are of vast dimensions." Animals are likewise exceptional: "The whole coast is abundantly stocked with hogs, sheep, goats, and all kinds of poultry, which propagate with astonishing rapidity." As to vegetables, Africa "affords very luxuriant pasturage—Millet, rice, maize, potatoes, yams, and a great variety of other excellent roots and vegetables, are cultivated on the coast with little trouble, and often in a profusion perfectly astonishing to an European." This image of unexploited fecundity is of course part of Wadström's mission to convince Europeans to settle the continent, and it must have been easy for many to dismiss it as exaggeration. To counter any such allegations, Wadström again blames the slave trade: "As the slave-ships never return directly to Europe, but proceed to the West Indies with their wretched cargoes," he points out, "it has never been in the interest of their owners to bring home much of the produce of Africa." Such African products as reach Europe, he argues, are actually cultivated in the West Indies. "What a strange inversion of natural order," he concludes, repeating the view he gave in his earlier book, "to exile from their native soil, both men and plants; the one to languish as slaves, and the other as exotics." Wadström's curious double movement is to denigrate existing European empires as unnatural while advocating further colonization along supposedly natural lines.[10]

Essay on Colonization unambiguously yokes together discourses of colonization, antislavery, and natural history and is thus a fitting example with which to conclude this study. Wadström may have been an extreme case, but abolitionist writers of the late eighteenth century certainly did bequeath a mixed legacy. They made good use of the evidence about slavery they found in natural histories, and they subverted the conventions of plantation management manuals to expose the corrupt and brutal reality the manuals attempted to justify. The slave trade, repeatedly decried as an unnatural business that perverted the laws of nature, was by their efforts made illegal. The scientific colonial gaze sought not only to understand, however, but also to control. While abolitionist writers were on the face of it sincere in wanting to set men and women free, they also believed implicitly that British ways of understanding, cataloguing, and finally exploiting the natural world were the best available, whether at home or abroad. Thus, colonial natural history, as mediated through abolitionist discourse, would in time give British readers the very best of reasons for supporting one of the very worst of projects in human history.

Notes

INTRODUCTION

1. Cobbett, *Parliamentary History,* vol. 28 (1789–91), 45, 48, 63–64.
2. Equiano, *Interesting Narrative,* 22, 192–93; Beatson, *Compassion the Duty and Dignity of Man,* 3; Booker, *Sermons,* 129; Stewart, *Reflections, Moral and Political,* 1:158. Clarkson's chest and its contents are on display at the Wisbech and Fenland Museum in Cambridgeshire. See Webster, "Collecting for the Cabinet of Freedom"; and Farrell, Unwin, and Walvin, "Thomas Clarkson's African Box," 308–13.
3. Roscoe, *Wrongs of Africa,* 1:13–14; More, *Slavery,* 10; Stockdale, *Letter from Percival Stockdale,* 3–4, 21–22; Priestley, *Sermon on the Subject of the Slave Trade,* 29.
4. The "decline theory"—that slavery was abolished only because it was no longer profitable—was put forward by Eric Williams in 1945 and became prominent in the 1960s. See Williams, *Capitalism and Slavery.* Williams's figures were challenged by Anstey in *The Atlantic Slave Trade,* 38–57, and later, and more comprehensively, by Drescher in *Econocide* and *Slavery to Freedom.* The main positions were anthologized in Bender, *The Antislavery Debate.*
5. Fairer, " 'Where fuming trees refresh the thirsty air,' " 202; Amussen, *Caribbean Exchanges,* 10; Bohls, *Slavery and the Politics of Place,* 8.
6. Allewaert, *Ariel's Ecology,* 18; Hosbey, Lloréns, and Roane, "Global Black Ecologies," 1; Ferdinand, *Decolonial Ecology,* 11, 38. For a lucid round-up of current thinking around the Anthopocene and Plantationocene, see Davis et al., "Anthropocene, Capitalocene, . . . Plantationocene?"
7. Williams, *Keywords,* 184.
8. Hobbes, *Leviathan,* ch. 13; Locke, *Two Treatises of Government,* 2:4, §24; Sharp, *Tract,* 3, 6, 9–14.

9. For early modern appropriations of classical biology, see Ogilvie, "Visions of Ancient Natural History." The classic study of the chain of being is Lovejoy, *Great Chain of Being*. The idea is explored in relation to slavery in Jordan, *White over Black*, 217–30. See also Carson, "The Great Chain of Being as an Ecological Idea." For a survey of the herbal from classical to early modern times, see Arber, *Herbals*.
10. Blome, *Description of the Island of Jamaica*; Rochefort, *Histoire naturelle*; Tryon, *Friendly Advice to the Gentlemen-Planters of the East and West Indies*.

CHAPTER 1. "THE CORD THAT BINDES UP ALL"

1. Ligon, *History*, 55 (all citations are taken from Ligon's original 1657 text); Steele, *Spectator* 11, 13 March 1711. For versions of the Inkle and Yarico story in several languages, see Price, *Inkle and Yarico Album*. For a recent comprehensive English-language collection, see Felsenstein, *English Trader, Indian Maid*. For discussion of Steele's adaptation in an Atlantic context, see Carey, " 'Accounts of Savage Nations.' "
2. Sandiford, *Theorizing a Colonial Caribbean-Atlantic Imaginary;* Stearns, *Science in the British Colonies*, 213; Grove, *Green Imperialism*, 67–71; Pluymers, *No Wood, No Kingdom*, 169. The plagiarized version is Clarke, *A True, and Faithful Account of the Four Chiefest Plantations of the English in America*. The reprint, with corrections and alterations, was published in London in 1673.
3. Stevenson, "Richard Ligon and the Theatre of Empire," 287; Ligon, *History*, 21. For biography, in addition to Stevenson, see Kupperman, introduction, 1–8.
4. Ligon, *History*, 2–3. Russets are brown autumn leaves while phyliamorts are dead leaves that have not yet fallen.
5. Ligon, *History*, 3.
6. Ligon, *History*, 8.
7. Ligon, *History*, 2, 12, 15, 22; Amussen, *Caribbean Exchanges*, 49; Datiles and Acevedo-Rodríguez, "*Caesalpinia pulcherrima* (peacock flower)." An asinigo is a "little ass."
8. Ligon, *History*, 6; Parrish, "Richard Ligon and the Atlantic Science of Commonwealths," 211.
9. Ligon, *History*, 3–4.
10. Ligon, *History*, 20.
11. Ligon, *History*, 20–21; Parrish, "Richard Ligon and the Atlantic Science of Commonwealths," 217; Stevenson, "Richard Ligon and the Theatre of Empire," 308.
12. Ligon, *History*, 24; Menard, *Sweet Negotiations*. The best general history of Barbados is Beckles, *A History of Barbados*. For an important study of the colony's beginnings, see Gragg, *Englishmen Transplanted*.
13. Watts, *Man's Influence*, 34, 38, 43, 45–46; Pluymers, *No Wood, No Kingdom*, 170–84. See also Watts, *The West Indies;* and Hollsten, "Controlling Nature."
14. Parrish, "Richard Ligon and the Atlantic Science of Commonwealths," 220; O'Connor, *Pineapple*, 17–41; Ligon, *History*, 84–85.
15. Ligon, *History*, 55, 57.
16. Ligon, *History*, 87; Overton, *Agricultural Revolution in England*, 16, 17.
17. Ligon, *History*, 87, 88. Compare Matthew 13:23, 7:26–27. Kuperman notes there are four withes in Barbados and considers the most likely candidate for Ligon's plant is

Cissus sicyoides. It is probable, however, that he had several similar plants in mind. See Kupperman's edition of Ligon, *True and Exact History,* 168*n.* For cane-holing, see Abbott, *Sugar,* 83–87; and Watts, "Origins of Barbadian Cane Hole Agriculture."

18. Ligon, *History,* 43, 48–49; Kupperman, ed., Ligon, *True and Exact History,* 100*n*119; Ramsay, *Essay,* 197–262.
19. Pluymers, *No Wood, No Kingdom,* 188; Bicknell and Day, *Dying Negro,* ix; Roscoe, *Wrongs of Africa,* 1:19. For the history of the banana and plantain in the Atlantic world, see Carney and Rosomoff, *In the Shadow of Slavery,* 34–43.
20. Ligon, *History,* 50.
21. Ligon, *History,* 52–53.
22. Ligon, *History,* 46–47.
23. Ligon, *History,* 48.
24. Ligon, *History,* 51; Amussen, *Caribbean Exchanges,* 63, 64; Morgan, *Laboring Women,* 14–15. Morgan refers to the general argument of Hall's *Things of Darkness.* Although Hall does not discuss Ligon directly, her overall claim is that medieval notions of light and dark became stable markers of white and black, and hence self and other, in early modern discourse. Hakluyt, *Principall Navigations,* 6 102, 101. For discussion, see Bartels, "Imperialist Beginnings," 526.
25. Ligon, *History,* 34, 36, 105.
26. Ramsay Papers, BL Add MS 27621, ff. 235–36, British Library, London; Sloane, *Voyage,* 1:7, 11, 21, 24, 27, 28, 34, 39 (and elsewhere); Benezet, *Some Historical Account,* 183; Oldmixon, *British Empire in America,* 1:xv; *Report of the Lords of the Committee of Council,* 271, 818; Steele, *Spectator,* 13 March 1711.

CHAPTER 2. "A VERY PERVERSE GENERATION OF PEOPLE"

1. *Some Modern Observations,* 4, 7–8.
2. The best recent biography is Delbourgo, *Collecting the World.* For the development of Jamaica as the wealthiest British colony, see Burnard, *Planters, Merchants, and Slaves.*
3. Ray, *Historia plantarum,* 1:40; Raven, *John Ray,* 306–7; Daston and Galison, *Objectivity,* 17; Sloane, *Voyage,* 1:preface.
4. Sloane, *Voyage,* 1:4–5, 20, 27; Ray, *The Ornithology of Francis Willughby,* 3:352–53.
5. Sloane, *Voyage,* 1:preface; Stearns, *Science in the British Colonies,* 243; Delbourgo, *Collecting the World,* 116.
6. Sloane, *Voyage,* 1:preface; Hooke, "A Discourse of Earthquakes in the Leeward Islands," 416–24. Ray's natural theology is explicitly set out in *The Wisdom of God.* For discussion of early evolutionary thought, see Thomson, *Before Darwin;* and Stott, *Darwin's Ghosts.*
7. Pope, *Essay on Man,* I, 289–94.
8. Tryon, *Friendly Advice,* 79–80. The classic account of slavery as sin is Davis, *Problem of Slavery in Western Culture,* ch. 3.
9. Sloane, *Voyage,* 1:33–34, xiv, 2:289.
10. Sloane, *Voyage,* 1:xlvi.
11. Ligon, *History,* 49; Sloane, *Voyage,* 1:xlvii, xlviii—xlix, l—li. The music scores, further reading, and a sound recording are available at Dubois, Garner, and Lingold, *Musical*

Passage. For discussion, see Rath, "African Music in Seventeenth-Century Jamaica," 700–726.

12. Sloane, *Voyage,* 1:liv, lvi, lvii; Delbourgo, "Slavery in the Cabinet of Curiosities," 15; Delbourgo, *Collecting the World,* 80.
13. Delbourgo, "Slavery in the Cabinet of Curiosities," 21; Delbourgo, *Collecting the World,* 79, 191.
14. Sloane, *Voyage,* 1:cxiv, cxli, cxxii.
15. Sloane, *Voyage,* 1:105, 103, 108–9, 111, 117, 136.
16. Sloane, *Voyage,* 2:ix. The anecdote was eventually published in Barham, *Hortus Americanus* in 1794
17. Clarkson, *History,* 1:55. For biography, see Dallett, "Griffith Hughes Dissected"; Clement, "Griffith Hughes"; and Shilstone, "Rev. Griffith Hughes."
18. *Monthly Review,* July 1750, 197–206, 200, 203; Stearns, *Science in the British Colonies,* 358, 359; Dallett, "Griffith Hughes Dissected," 23–24.
19. Hughes, *Natural History,* 75, 237, 262. The bird is probably the yellow warbler (*Setophaga petechia*) and the plant probably the black maidenhair fern (*Adiantum capillus-veneris*). A plausible candidate for the lobster is the Spanish slipper lobster (*Scyllarides aequinoctialis*).
20. Hughes, *Natural History,* iii—iv, 303. Hughes describes the flying fish being "pursued by *Dolphins,* or other voracious Fish." He is referring to an actual fish, the common dolphinfish or dorado (*Coryphaena hippurus*), which in the United States is sometimes known by its Hawaiian name "mahi-mahi."
21. Hughes, *Natural History,* 51, 67, 88–89, 123–24.
22. Hughes, *Natural History,* 251–52.
23. Hughes, *Natural History,* 13, 15.
24. Hughes, *Natural History,* 14, 16. For skin color and anatomy, see Wheeler, *Complexion of Race,* 26.
25. Hughes, *Natural History,* 17–18.
26. Hughes, *Natural History,* 19.

CHAPTER 3. "NEGROES, CATTLE, MULES, AND HORSES"

1. Silva, "Georgic Fantasies," 127.
2. Fisher, "The Master Should Know More," 315.
3. Mylander, "Early Modern 'How-To' Books," 125.
4. Zacek, "Yeomen or 'Grandy-Men?' " 5 and *Guide to the . . . Papers of Samuel Martin,* 4; Schaw, *Journal of a Lady of Quality,* 103–4; Sheridan, "Samuel Martin"; Martin, *Essay upon Plantership,* 5th ed., ix. The correspondence is at the British Library in London. See Zacek, *Guide to the . . . Papers of Samuel Martin.* Turnbull, *Letters,* "Advertisement." The ruling in the Somerset case was ambiguous, preventing slave owners from compelling their slaves to leave the country, but not technically making slavery illegal. See Wise, *Though the Heavens May Fall.* A useful roundup of later proslavery plantation management manuals can be found in Dumas, *Proslavery Britain,* 81–84.
5. Martin, *Essay upon Plantership,* 2nd ed., 9–11.

6. Martin, *Essay upon Plantership,* 2nd ed., vii, 21, 32–33, 12–14. For the identification of the insect, see Kevan, "Mid-Eighteenth-Century Entomology." Hughes, *Natural History,* 245–46; Foy, "The Convention of Georgic Circumlocution," 476.
7. Martin, *Essay upon Plantership,* 2nd ed., 14, 18, 30–31; Boulukos, *Grateful Slave,* 93. The idea of the plantation as a machine has been explored by Benitez-Rojo in *The Repeating Island,* 8–9 and passim.
8. This section builds on some of the ideas explored in my essay "James Grainger's *The Sugar-Cane* and Naturalists' Georgic."
9. Mackenzie, *Julia de Roubigné,* 2:41. For discussion, see Carey, *British Abolitionism,* 63–67. Grainger, *Sugar-Cane,* 3:31–38, preface. Gilmore suggests that Grainger may also have intended "M" to refer to his friend the planter Daniel Matthew (1718–77). See Gilmore, *Poetics,* 264–65.
10. Turnbull, *Letters,* "Advertisement"; O'Brien, "Imperial Georgic," 160–79; Grainger, *Sugar-Cane,* preface; Boswell, *The Life of Samuel Johnson,* 21 March 1776, 2:453–54, discussed at 533; Gilmore, *Poetics,* 2–21. Chalker, *The English Georgic,* 60; Randhawa, "The Inhospitable Muse," 68.
11. Gilmore, *Poetics,* 29; Ellis, " 'Incessant Labour,' " 49; Foy, "Grainger and the 'Sordid Master.' " The classic general study of georgic is Chalker, *The English Georgic.* For the mid-eighteenth-century revival, see Bucknell, "The Mid-Eighteenth-Century Georgic."
12. Grainger, *Sugar-Cane,* 1:218–27; "O'Brien, "Imperial Georgic," 174; Gilmore, *Poetics,* 225; Loar, "Georgic Assemblies," 246; Senior, *The Caribbean and the Medical Imagination,* 53. "Stale" is urine, particularly of horses and cattle.
13. Grainger, *Sugar-Cane,* 3:623–29, 635–48; Irlam, " 'Wish You Were Here,' " 382, 379. "For naught is useless made" is from Philips, *Cyder,* I, 98. See Gilmore, *Poetics,* 285.
14. Grainger, *Sugar-Cane,* 2"201–5, 246–52, 261–62, 205*n.* Allewaert, "Insect Poetics," 302. For more on Grainger as naturalist, see Carey, "James Grainger's *The Sugar-Cane* and Naturalists' Georgic."
15. Rusert, "Plantation Ecologies," 344, 348–49; Grainger, *Sugar-Cane,* 1:4, 268; Chalker, *The English Georgic,* 56. Gilmore notes that where Grainger wrote of Africans, the equivalent line in Virgil's *Georgics* was "The care of sheep, of oxen, and of kine" and that Virgil contrasts orderly planting of trees with battle lines. See Gilmore, *Poetics,* 215, 227.
16. Grainger, *Sugar-Cane,* 3"98, 100, 145–46; Douglass, *Narrative of the Life of Frederick Douglass,* 58.
17. Grainger, *Sugar-Cane,* 4:211, 421, 424–25, 232–43; Egan, "The 'Long'd-for Aera,' " 196; Thomas, "Doctoring Ideology," 79; Sandiford, *Cultural Politics of Sugar,* 86; Ellis, " 'Incessant Labour,' " 57; Silva, "Georgic Fantasies," 129.
18. Silva, "Georgic Fantasies," 129; Grainger, *Sugar-Cane,* 4:206–11.
19. The ruling was made by William Murray, 1st Earl of Mansfield (1705–93) in the case of James Somerset (c. 1741—after 1772) on 22 June 1772. The case is discussed at length in Wise, *Though the Heavens May Fall.*
20. Long, *History,* 1:2; *Monthly Review,* July–December 1774, 129–36, 431–41. Donaldson (d. 1780) had been secretary to the government of Jamaica. See Nangle, *Monthly Review,* 125, 14.

21. Long, *History,* 2:123–28.
22. Long, *History,* 2:92–94, 101.
23. Long, *History,* 2:134–37.
24. Long, *History,* 1:5; Wise, *Though the Heavens May Fall,* 182.
25. Long, *History,* 2:336, 352–53, 364–65.
26. Wheeler, *Complexion of Race,* 211; Fryer, *Staying Power,* 158; Swaminathan, *Debating the Slave Trade,* 165; Long, *History,* 2:372; Estwick, *Considerations,* 74; White, *Account,* 1, 42. For Long's environmental politics, exploring his use of landscape aesthetics "in the service of early scientific racism," see Bohls, "The Gentleman Planter and the Metropole."
27. Long, *History,* 2:374–75, 503.
28. Loar, "Georgic Assemblies," 243–44.

CHAPTER 4. "THE PURCHASE OF SLAVES, TEETH AND DUST"

1. For a useful summary of early English travelers to West Africa, see Hair and Law, "The English in West Africa to 1700." For facsimiles of primary texts relating to the Royal African Company, with useful introductions, see Morgan, *The British Atlantic Slave Trade.* For the history of slavery in Africa, see Lovejoy, *Transformation in Slavery.* For Africa in the literary culture of early modern England, see Hall, *Things of Darkness.*
2. Murphy, *Captivity's Collections,* 11; Day, " 'Honour to Our Nation,' " 77.
3. The best general history of the slave trade in this period remains Thomas, *The Slave Trade.* For the political history of the Royal African Company, see Pettigrew, *Freedom's Debt.* For the social history of the Middle Passage, see Rediker, *The Slave Ship.*
4. Jerom Merolla da Sorrento, *A Voyage to Congo, and Several Other Countries Chiefly in Southern-Africk,* in *Churchill's Collection,* 1:592–695, 606–7.
5. Merolla, *Voyage to Congo,* 634–40; Gray, *Black Christians,* 20–31.
6. Merolla, *Voyage to Congo,* 610–11; *Astley's Collection,* 1:v.
7. Thomas Phillips, *A Journal of a Voyage Made in the Hannibal of London, Ann. 1693, 1694, from England to Cape Monseradoe in Africa,* in *Churchill's Collection,* 6:172–239, 173. This was not apparently printed elsewhere. Law, "Jean Barbot as a Source for the Slave Coast of West Africa," 155; Barbot, *Barbot on Guinea,* 1:xxi. Hair's important publication offers a composite of Barbot's original 1688 manuscript in a modern translation, with excerpts from the 1732 text in *Churchill's Collection,* while his introduction and notes offer detailed analysis of the composition of the text, its sources, and Barbot's biography. Contemporary readers would, however, have encountered only the 1732 edition.
8. Jean Barbot, *A Description of the Coasts of North and South-Guinea,* in *Churchill's Collection,* 5:11, 3.
9. Barbot, *Description,* 27–28; Barbot, *Barbot on Guinea,* 1:70.
10. Barbot, *Description,* 73, 112–17, 206–27, 128; Barbot, *Barbot on Guinea,* 1:156. The Quoja passage does not occur in the 1688 manuscript, but I have been unable to identify another printed source for it. For the Gold Coast passage, compare Bosman, *Description,* 234–83.

11. Barbot, *Description,* 102–6. The passages appear to conflate a number of sources. See Barbot, *Barbot on Guinea,* 1:186–88.
12. Barbot, *Description,* 77–78.
13. Barbot, *Description,* 270–72, 546–48. The latter passage does not appear in the 1688 manuscript. For Quaker fear of enslavement at the hands of Muslims, and its application to the Golden Rule, see Carey, *From Peace to Freedom,* 78–79, 94.
14. *Astley's Collection,* 1:v, vii. For biography, see "John Green, Geographer," Crone, "A Note on Bradock Mead"; and Crone, "Further Notes on Bradock Mead." The Dublin-born Mead is revealed as "a womanizer who perhaps supported himself by gambling" and who took part in a failed "plot to cheat a twelve-year-old Irish heiress of her family's property. . . . He eventually committed suicide, by leaping out of a third-storey window, in 1757."
15. Chambers, *Cyclopædia,* 2:495–97, 889; Benezet, *Complete Antislavery Writings,* 268.
16. Atkins, *Voyage,* 39, 1 (Atkin's original discussion of race appears in *The Navy-Surgeon,* appendix, 23–24); Clarkson, *History,* 1:52; *Astley's Collection,* 2:446, v.
17. Thomas, *The Slave Trade,* 339; *Astley's Collection,* 2:316–17. Compare Atkins, *Voyage,* 39–42. Sloane discusses the manatee-hide whip in *Voyage,* 1:lvii, discussed in chapter 2 of this book.
18. *Astley's Collection,* 2:449, 506. For details of the voyage of the *Robert,* see the https://slavevoyages.org/ database, voyage ID 16303, and compare Atkins, *Voyage,* 71–73. Atkin's antislavery sentiment is at *Voyage,* 149–50, 178. See also 61–62 and compare Snelgrave, *New Account,* 157–61.
19. *Astley's Collection,* 2:242. Compare Moore, *Travels into the Inland Parts of Africa,* 42.
20. *Astley's Collection,* 2:255, 267, 268. Compare Labat, *Nouvelle relation de l'Afrique occidentale,* iv, 351.
21. *Astley's Collection,* 2:324.
22. *Astley's Collection,* 2:359 and fig. xix (facing 143). Compare Barbot, *Description,* 132 and plate F.
23. The date palm is *Phoenix dactylifera.* The coconut palm is *Cocos nucifera.* The identity of the "Areka" is unclear since the five species of *Areca,* including the areca palm (*A. catechu*) are Asian-Pacific trees, not found in West Africa. The "Cypress palm" is probably the African fan palm (*Borassus aethiopum*), one of several trees that yield palm wine. "Manjok" is manioc, another name for cassava (*Manihot esculenta*) "Patatas" appear to be several species of yam or other tubers. "Great millet or maez" is probably maize, while "small millet" might refer to many species of grains, but almost certainly not *Panicum sumatrense,* the Asian species currently known as little millet. "Kuskus" (normally "couscous") is not a plant but a preparation of wheat or other grains, which Green acknowledges.
24. *Astley's Collection,* 2:34–48, 351–52, 349. For the elephant passages, compare Bosman, *Description,* 242–45.
25. *Astley's Collection,* 2:362 (compare Bosman, *Description,* 281–83, and a considerably toned-down version also appears in Barbot, *Description,* 226); Rediker, *The Slave Ship,* 39, 40; Equiano, *Interesting Narrative,* 47; Jamieson, *The Sorrows of Slavery,* 33; Shelley, "To S. and C.," in *The Major Works,* 443–44; Heyrick, *Immediate, Not Gradual, Abolition,* 9.

26. Adanson, *Voyage,* iv–v. For "*auri sacra fames*" ("hunger for gold") see Virgil, *Aeneid,* 3:57.
27. Adanson, *Voyage,* 126, 96–98, 76–77.
28. Adanson, *Voyage,* 253, 152–53, 251–52, 159–62, 284. The bird is probably either the golden sparrow (*Passer luteus*), the village weaver, or the black-headed weaver (*Ploceus melanocephalus*).
29. Adanson, *Voyage,* 150–61, 155, 105, 178–79, 249–50, 164, 300–301. Many of the organisms he mentions are vague and hard to identify. The term "curlew," for example, might describe any number of long-billed waders, while terms such as "duck" or "rock fish" are equally imprecise. The shellfish is probably the West African mangrove oyster (*Crassostrea tulipa*).
30. Adanson, *Voyage,* 43, 75, 54, 58–59, 68, 74.
31. Adanson, *Voyage,* 148*n*.

CHAPTER 5. "THE GROANS, THE DYING GROANS, OF THIS DEEPLY AFFLICTED AND OPPRESSED PEOPLE"

1. Major biographies of Benezet include Brookes, *Friend Anthony Benezet;* and Jackson, *Let This Voice Be Heard.* For a detailed account of the rise of Quaker abolitionism, including discussion of Benezet (211–14), see Carey, *From Peace to Freedom.* See also Carey and Plank, *Quakers and Abolition;* Drake, *Quakers and Slavery in America;* Rossignol and Van Ruymbeke, *The Atlantic World of Anthony Benezet;* Soderlund, *Quakers and Slavery.*
2. Benezet, *Complete Antislavery Writings,* 113; Sassi, "Africans in the Quaker Image," 99; Brown, *Moral Capital,* 137; Herschthal, *Science of Abolition,* 97–98. See also Sassi, "With a Little Help from the Friends."
3. Benezet, *Some Historical Account,* 3; Jackson, *Let This Voice Be Heard,* 72, 79; Sassi, "Africans in the Quaker Image," 99, 109, 105. For earlier discussions of Benezet's representation of Africa, see, in particular, Davis, *Problem of Slavery in Western Culture,* 467; and Rice, *Rise and Fall of Black Slavery,* 17, 200.
4. *Astley's Collection,* 1:vii.
5. Benezet, *Some Historical Account,* 14–15 (compare Adanson, *Voyage,* 164–65), 25 (compare *Churchill's Collection,* 5:154, 28 (compare *Astley's Collection,* 2:488), 39, 18, 35; Snelgrave, *New Account,* 3.
6. Benezet, *Some Historical Account,* 14–16; Adanson, *Voyage,* 54, 159–62. Although he was a native French speaker, Benezet appears to be using the English translation of Adanson's voyage with minor alterations.
7. Jackson, *Let This Voice Be Heard,* 80; Sassi, "Africans in the Quaker Image," 104, 129; Benezet, *Some Historical Account,* iii—iv, 60–61. Compare Smith, *A New Voyage to Guinea,* 266. In this case, Benezet is quoting from the original text rather than from *Astley's Collection.*
8. Benezet, *Some Historical Account,* 23–24. Compare Bosman, *Description,* 16–17, 47. In his footnotes, Benezet gives page references to the 1721 2nd edition of Bosman's *Description.*
9. Labat, *Nouvelle relation;* Benezet, *Some Historical Account,* 7. Compare *Astley's Collection,* 2:49, 86.

10. Benezet, *Some Historical Account,* 121; Coffey, " 'I Was an Eye-witness.' "
11. Benezet, *Some Historical Account,* 85–86; Hughes, *Natural History,* 14.
12. Benezet, *Some Historical Account,* 86–87. Compare Burke, *Account of the European Settlements,* 2:120–21; and Jefferys, *Natural and Civil History,* 2:186. For Benezet and sensibility, see Carey, "Anthony Benezet, Antislavery Rhetoric and the Age of Sensibility."
13. Benezet, *Some Historical Account,* 89–90. Compare Sloane, *Voyage,* 1:lvii.
14. Benezet, *Some Historical Account,* 92–94.
15. Benezet, *Some Historical Account,* 142, 183. Compare Ligon, *History,* 27.
16. Brendlinger, *To Be Silent,* 18; Davis, *Problem of Slavery in the Age of Revolution,* ch. 5, passim.

CHAPTER 6. "AN UNNATURAL STATE OF OPPRESSION"

1. For a biography of Sharp in his family context, see Grant, *The Good Sharps.*
2. Shyllon, *James Ramsay,* remains the only substantial biography of Ramsay. Equiano, *Interesting Narrative,* 86; Clarkson, *History,* 1:103. For extended discussion of the Ramsay-Tobin dispute, see Carey, *British Abolitionism,* 119–24; Swaminathan, *Debating the Slave Trade,* 148–53; and Swaminathan, "(Re)Defining Mastery." The Zong massacre is examined in Baucom, *Specters of the Atlantic;* and Walvin, *The Zong.*
3. Ramsay, *Essay,* 231.
4. *European Magazine* 13 (1788): 75–76, discussed in Shyllon, *James Ramsay,* 130–31; Smith, *A Letter from Capt. J. S. Smith,* v–vi; Ramsay, *Reply to the Personal Invectives,* 60.
5. Ramsay Papers: Bodleian MSS.Brit.Emp.s.2, Bodleian Library, Oxford, and BL Add. MS 27621, ff. 235–36, 1–7, 192, 204–5, British Library, London; Ramsay, *Essay,* 259; *Essai sur l'histoire naturelle de St. Domingue* (Paris, 1776), authorised by Michel Adanson as the work of "Le P. Nicolson" (Father Nicolson). Henry Smeathman (1742–86) was an entomologist who was behind the Sierra Leone colonization scheme to resettle London's poor Black population. See Coleman, *Henry Smeathman.*
6. Ramsay, *Essay,* 1, 2, 19.
7. Ramsay, *Essay,* iii, 69–70, 73; Martin, *Essay upon Plantership,* 2nd ed., 15.
8. Ramsay Papers, BL Add. MS 27621, f. 178, British Library, London; Martin Papers, BL Add. MS 41350, f. 26, British Library, London.
9. Martin, *Essay upon Plantership,* 2nd ed., 21; Ramsay, *Essay,* 73.
10. Martin, *Essay upon Plantership,* 2nd ed., 16–18.
11. Martin, *Essay upon Plantership,* 2nd ed., 12–14, 9; Ramsay, *Essay,* 78–80.
12. Ramsay, *Essay,* 269. He articulated a more vigorous antislavery position later the same year in *Inquiry into the Effects of Putting a Stop to the African Slave trade.*
13. Hochschild, *Bury the Chains,* 91; Clarkson, *Essay,* 66. The most recent biography is Wilson, *Thomas Clarkson.* For the composition of the essay, see Clarkson, *History,* 1:207; Wilson, *Thomas Clarkson,* 9–15.
14. Clarkson, *Essay,* vi–ix, 43, 112–13 (compare Benezet, *Some Historical Account,* 125), xxv.
15. Clarkson, *Essay,* 41, 50–53. Compare Sparrman, *A Voyage to the Cape of Good Hope,* 2:143. The geographer Thomas Jefferys (c. 1719–71), wrote many works on America

and the Caribbean, including *A General Topography of North America and the West Indies.*

16. Clarkson, *Essay,* 117, 128, 142, xi–xii; Carretta, "Equiano's Paradise Lost," 82–83; Francklyn, *An Answer to the Rev. Mr. Clarkson's Essay,* 192.
17. Clarkson, *Essay,* 164, 187–88, 190; Sinha, *The Slave's Cause,* 99; Hochschild, *Bury the Chains,* 92.
18. Clarkson, *Essay,* 176; Carretta, *Equiano, the African,* 335.
19. In addition to *The Interesting Narrative* itself, the standard biography is Carretta, *Equiano, the African,* in which Carretta makes the case that Equiano's birthplace was South Carolina, not Africa (see especially xiv, 8, 319–20). For extended discussion of the birthplace controversy, see Carey, "Olaudah Equiano: African or American?" I am following the widespread convention of using the author's African name, Olaudah Equiano, although throughout his life he used the name Gustavus Vassa in everyday contexts.
20. Bohls, *Slavery and the Politics of Place,* 142; Hosbey, Lloréns, and Roane, "Global Black Ecologies," 3.
21. Equiano, *Interesting Narrative,* 18, 23, 192, 193; Carretta, *Equiano, the African,* xiv; Lovejoy, "Autobiography and Memory," 336. Acholonu, in *The Igbo Roots of Olaudah Equiano,* offers an identification of Equiano's village based on linguistic analysis and oral history, but her proposed location has not been universally accepted.
22. Equiano, *Interesting Narrative,* 28.
23. Shuker, *Extraordinary Animals,* 31–45. For West African snakes more generally, see Chippaux and Jackson, *Snakes of Central and Western Africa.* Although I have been unable to find evidence of crowing-snake stories being told in West Africa, it is important to remember that the absence of evidence is not, in itself, evidence of absence.
24. Equiano, *Interesting Narrative,* 39, 47, 48; Stein, "Who's Afraid of Cannibals?" 99, Woodard, Joyce, and McBride, *Delectable Negro,* 37.
25. Equiano, *Interesting Narrative,* 109, 125.

CHAPTER 7. "BUT SAY, WHENCE FIRST TH'UNNATURAL TRADE AROSE?"

1. Oldfield, *Popular Politics,* 1; Wordsworth, *The Prelude* (1805), 10:212; Basker, *Amazing Grace,* xxxiv. For abolition and sensibility, see Ahern, *Affect and Abolition*; and Carey, *British Abolitionism.*
2. *Morning Chronicle and London Advertiser,* 28 May 1773. The report also appeared in the *General Evening Post,* 25–27 May 1773, and *Lloyd's Evening Post,* 26–28 May 1773. For the Somerset case, see Wise, *Though the Heavens May Fall.*
3. Moore, *How to Create the Perfect Wife,* 219–25. For a general biography, see Rowland, *The Life and Times of Thomas Day.* On Day and the Lunar Society, see Uglow, *The Lunar Men.* For *Dying Negro* and sentiment, see Carey, *British Abolitionism,* 75–84; for the poem as a representation of suicide, see Snyder, *The Power to Die,* 121–39.
4. Bicknell and Day, *Dying Negro,* 10–12 (these lines and the footnotes, both of which are Day's, are unchanged across the various editions of the poem; the Adanson quote

in Benezet's *Some Historical Account* is at 93, the Smith quote is at 127, and the Barbot quote at 164); Milton, *Paradise Lost,* 9:501–4; Pope, "An Epistle from Mr. Pope to Dr. Arbuthnot," 330–31.

5. Bicknell and Day, *Dying Negro,* 4, 5, 8, 6, 7.
6. Bicknell and Day, *Dying Negro,* 16, 17.
7. Bicknell and Day, *Dying Negro,* 21, 23–24.
8. For biography, see Baines's introduction to *The Collected Writings of Edward Rushton*; and Dellarosa, *Talking Revolution.*
9. Crandall, " 'The Great Measur'd by the Less,' " 976; Dellarosa, *Talking Revolution,* 143, 164, 162; Rushton, *Eclogues,* 15, 29, 26. Compare *Astley's Collection,* 2:55, 348; Sloane, *Voyage,* 1:lvii; and Benezet, *Some Historical Account,* 155.
10. *English Review,* October 1787, 315; *Monthly Review,* October 1787, 283; *Critical Review,* December 1787, 434–35.
11. Curran, *Poetic Form,* 96; Dellarosa, *Talking Revolution,* 144; Pierrot, "Sable Warriors and Neglected Tars," 126–27; Dellarosa, *Talking Revolution,* 142–43, 145; Rushton, *Collected Writings,* 64–73.
12. Rushton, *Collected Writings,* 234–35.
13. Dellarosa, *Talking Revolution,* 161; Rushton, *Eclogues,* 1.
14. Rushton, *Eclogues,* 7, 10, 11, 27–28; Chatterton, "Heccar and Gaira"; Brown, "A Description of the *Exocœtus Volitans,* or *Flying Fish*"; Sloane, *Voyage,* 1:27.
15. Rushton, *Eclogues,* 13–19.
16. Pierrot, "*Droit du Seigneur,*" 15–16; Rushton, *Eclogues,* 21, 22, 24; Pierrot, *Black Avenger,* 75; Chatterton, "Heccar and Gaira."
17. Clarkson, *History,* 1:279–83; Roscoe, *Wrongs of Africa,* 2:6.
18. *Monthly Review,* February 1788, 137; *English Review,* October 1787, 261; Basker, *Amazing Grace,* 196, 197; Wood, *Poetry of Slavery,* 53; Moody, *Persistence of Memory,* 134.
19. Stott, *Hannah More,* 86–95; Andrews, *Ann Yearsley and Hannah More,* 85–93; Basker, *Amazing Grace,* 335; More, *Slavery,* 11; *Monthly Review,* March 1788, 246; *Critical Review,* March 1788, 226; Wood, *Poetry of Slavery,* 100. For the relationship between More and Yearsley, see Andrews, *Ann Yearsley and Hannah More.*
20. Clarkson, *History,* 2:188–91; Cowper, *Letters and Prose Writings,* 3:130–31.
21. More, *Slavery,* 11, 10; Roscoe, *Wrongs of Africa,* 1:13–14 (compare Milton, *Paradise Lost,* 1:27–33), 19, 14–15, 29; Milton, *Paradise Lost,* 2:686–87, 4:121–22.
22. More, *Slavery,* 1–2, 4–6 (compare Ramsay, *Essay,* 249–53); Roscoe, *Wrongs of Africa,* 1:ii, vi. For the collaboration between Roscoe and Currie, see Krawczyk, "Mediating Abolition."
23. More, *Slavery,* 20; Roscoe, *Wrongs of Africa,* 2:43, 1:9.
24. Cowper, "The Negro's Complaint," 13–14, 33–40; Roscoe, *Wrongs of Africa,* 1:7, 8.
25. Roscoe, *Wrongs of Africa,* 1:9; Cowper, "The Negro's Complaint," 17–20.
26. Roscoe, *Wrongs of Africa,* 1:4, 5, 9, 12, 13–14, 2:11 (compare Milton, *Paradise Lost,* 1:59–69); Milton, *Paradise Lost,* 4:160–61.
27. More, *Slavery,* 19–20, 10, 13; Stott, *Hannah More,* 90.
28. Cowper, "The Negro's Complaint," 55–56; Ramsay, *Essay,* 19.

CONCLUSION

1. Hansard, 23 February 1807.
2. Smith, *A Letter to William Wilberforce,* 11, 27, 29; *Gleanings in Africa,* 72, 73; Mant, *The Slave,* 18, 16, 17, 24. Compare Milton, *Paradise Lost,* 1:743.
3. Bales, *Disposable People,* 121–22.
4. Smith, *Wealth of Nations,* 1:99; Benezet, *Some Historical Account,* 3, 18, 144.
5. Clarkson, *Essay,* 142, x–xii; Equiano, *Interesting Narrative,* 23, 192–93; *Parliamentary History* 28 (1789–91): 63–64.
6. Gascoigne, *Science in the Service of Empire,* 180; Mackay, *In the Wake of Cook,* 123—43; Gascoigne, *Joseph Banks,* 39–44. For Smeathman, see especially Coleman, *Henry Smeathman*; and Herschthal, *Science of Abolition,* 98–111.
7. Herschthal, *Science of Abolition,* 97.
8. Wadström, *Observations,* iv, 42–43, 60–61.
9. Wadström, *Essay on Colonization,* 1:iii–iv, 21–22.
10. Wadström, *Essay on Colonization,* 1:15, 27, 34, 35, 33.

Bibliography

Abbott, Elizabeth. *Sugar: A Bittersweet History.* London: Duckworth, 2009.

Acholonu, Catherine. *The Igbo Roots of Olaudah Equiano.* Owerri, Nigeria: Afa, 1989.

Adanson, Michel. *Histoire naturelle du Sénégal. Coquillages. Avec la relation abrégée d'un voyage fait en ce pays, pendant les années 1749, 50, 51, 52 & 53.* Paris, 1757.

———. *A Voyage to Senegal, the Isle of Goree, and the River Gambia. By M. Adanson . . . Translated from the French. With Notes by an English Gentleman, Who Resided Some Time in That Country.* London, 1759.

Addison, Joseph, and Richard Steele. *The Spectator.* 5 vols. Edited by Donald F. Bond. Oxford: Clarendon, 1965.

Ahern, Stephen, ed. *Affect and Abolition in the Anglo-Atlantic: 1770–1830.* Farnham, UK: Ashgate, 2013.

Allewaert, Monique. *Ariel's Ecologies: Plantations, Personhood, and Colonialism in the American Tropics.* Minneapolis: University of Minnesota Press, 2013.

———. "Insect Poetics: James Grainger, Personification, and Enlightenments Not Taken." *Early American Literature* 52:2 (2017): 299–332.

Amussen, Susan Dwyer. *Caribbean Exchanges: Slavery and the Transformation of English Society, 1640–1700.* Chapel Hill: University of North Carolina Press, 2007.

Andrews, Kerri. *Ann Yearsley and Hannah More, Patronage and Poetry: The Story of a Literary Relationship.* London: Pickering and Chatto, 2013.

Anstey, Roger. *The Atlantic Slave Trade and British Abolition, 1760–1810.* London: Macmillan, 1975.

Arber, Agnes. *Herbals: Their Origin and Evolution; A Chapter in the History of Botany, 1470–1670*, 2nd ed. Cambridge: Cambridge University Press, 1938.

Aristotle. *History of Animals*. 3 vols. Translated by A. L. Peck and D. M. Balme. Cambridge, MA: Harvard University Press, 1989–91.

———. *Meteorologica*. Translated by Henry D. P. Lee. London: William Heinemann, 1952.

Astley's Collection. See *A New General Collection of Voyages and Travels*.

Atkins, John. *The Navy-Surgeon: or, A Practical System of Surgery*. London, 1735.

———. *A Voyage to Guinea, Brasil, and the West-Indies; in His Majesty's Ships, the* Swallow *and* Weymouth. London, 1735.

Bales, Kevin. *Disposable People: New Slavery in the Global Economy*. 2nd ed. Berkeley: University of California Press, 2012.

Barbot, Jean. *Barbot on Guinea: The Writings of Jean Barbot on West Africa, 1688–1712*. 2 vols. Edited by Paul Hair. London: Hakluyt Society, 1992.

Barham, Henry. *Hortus Americanus: Containing an Account of the Trees, Shrubs, and Other Vegetable Productions, of South-America and the West-India Islands, and Particularly of the Island of Jamaica; Interspersed with Many Curious and Useful Observations, Respecting Their Uses in Medicine, Diet, and Mechanics*. Kingston, Jamaica, 1794.

Bartels, Emily C. "Imperialist Beginnings: Richard Hakluyt and the Construction of Africa." *Criticism* 34 (1992): 517–38.

Basker, James G., ed. *Amazing Grace: An Anthology of Poems about Slavery, 1660–1810*. New Haven, CT: Yale University Press, 2002.

Baucom, Ian. *Specters of the Atlantic: Finance Capital, Slavery, and the Philosophy of History*. Durham, NC: Duke University Press, 2005.

Beatson, John. *Compassion the Duty and Dignity of Man; and Cruelty the Disgrace of His Nature*. Hull, UK, 1789.

Beckles, Hilary McD. *A History of Barbados: From Amerindian Settlement to Caribbean Single Market*. 2nd ed. Cambridge: Cambridge University Press, 2006.

Bender, Thomas, ed. *The Antislavery Debate: Capitalism and Abolitionism as a Problem in Historical Interpretation*. Berkeley: University of California Press, 1992.

Benezet, Anthony. *The Complete Antislavery Writings of Anthony Benezet, 1754–1783*. Edited by David L. Crosby. Baton Rouge: Louisiana State University Press, 2013.

———. *Some Historical Account of Guinea, Its Situation, Produce, and the General Disposition of Its Inhabitants. With an Inquiry into the Rise and Progress of the Slave Trade, Its Nature, and Lamentable Effects*. London, 1772.

Benitez-Rojo, Antonio. *The Repeating Island: The Caribbean and the Postmodern Perspective*. 2nd ed. Translated by James E. Maraniss. Durham, NC: Duke University Press, 1997.

Bicknell, John, and Thomas Day. *The Dying Negro, a Poem.* 3rd ed. London, 1775.

Blome, Richard. *A Description of the Island of Jamaica; with the Other Isles and Territories in America, to Which the English Are Related.* London, 1672.

Bohls, Elizabeth A. "The Gentleman Planter and the Metropole: Long's *History of Jamaica* (1774)." In *The Country and the City Revisited: England and the Politics of Culture, 1550–1850,* edited by Gerald MacLean, Donna Landry, and Joseph P. Ward, 180–96. Cambridge: Cambridge University Press, 1999.

———. *Slavery and the Politics of Place: Representing the Colonial Caribbean, 1770–1833.* Cambridge: Cambridge University Press, 2014.

Booker, Luke. *Sermons on Various Subjects, Intended to Promote Christian Knowledge and Human Happiness.* Dudley, UK, 1793.

Bosman, Willem. *A New and Accurate Description of the Coast of Guinea, Divided into the Gold, the Slave, and the Ivory Coasts.* London, 1705.

Boswell, James. *The Life of Samuel Johnson, LL.D.* Edited by G. B. Hill. Revised by L. F. Powell. Oxford: Clarendon, 1964.

Boulukos, George. *The Grateful Slave: The Emergence of Race in Eighteenth-Century British and American Culture.* Cambridge: Cambridge University Press, 2008.

Brendlinger, Irv A. *To Be Silent . . . Would Be Criminal: The Antislavery Influence and Writings of Anthony Benezet.* Lanham, MD: Scarecrow, 2007.

Brookes, George S. *Friend Anthony Benezet.* Philadelphia: University of Pennsylvania Press, 1937.

Brown, Christopher Leslie. *Moral Capital: Foundations of British Abolitionism.* Chapel Hill: University of North Carolina Press, 2006.

Brown, Thomas. "A Description of the *Exocœtus Volitans,* or *Flying Fish.*" *Philosophical Transactions* 68 (1778): 791–800.

Browne, Patrick. *A Civil and Natural History of Jamaica.* London, 1756.

Bucknell, Clare. "The Mid-Eighteenth-Century Georgic and Agricultural Improvement." *Journal for Eighteenth-Century Studies* 36:3 (2013): 335–52.

Burke, William. *An Account of the European Settlements in America.* 2 vols. London, 1757.

Burnard, Trevor. *Planters, Merchants, and Slaves: Plantation Societies in British America, 1650–1820.* Chicago: University of Chicago Press, 2015.

Carey, Brycchan. " 'Accounts of Savage Nations': The *Spectator* and the Americas." In *Uncommon Reflections: Emerging Discourses in "The Spectator,"* edited by Don Newman, 129–49. Newark: University of Delaware Press, 2005.

———. "Anthony Benezet, Antislavery Rhetoric and the Age of Sensibility." *Quaker Studies* 21:2 (2016): 7–24.

———. *British Abolitionism and the Rhetoric of Sensibility: Writing, Sentiment, and Slavery, 1760–1807.* Basingstoke: Palgrave Macmillan, 2005.

———. *From Peace to Freedom: Quaker Rhetoric and the Birth of American Antislavery, 1657–1761.* New Haven, CT: Yale University Press, 2012.

———. "James Grainger's *The Sugar-Cane* and Naturalists' Georgic." In *Georgic Literature and the Environment: Working Land, Reworking Genre,* edited by Sue Edney and Tess Somervell, 73–88. London: Routledge, 2022.

———. "Olaudah Equiano: African or American?" *1650–1850: Ideas, Æsthetics, and Inquiries in the Early Modern Era* 17 (2010): 229–48.

Carey, Brycchan, Markman Ellis, and Sara Salih, eds. *Discourses of Slavery and Abolition: Britain and Its Colonies, 1760–1838.* Basingstoke: Palgrave Macmillan, 2004.

Carey, Brycchan, and Geoffrey Plank, eds. *Quakers and Abolition.* Champaign: University of Illinois Press, 2014.

Carney, Judith A., and Richard Nicholas Rosomoff. *In the Shadow of Slavery: Africa's Botanical Legacy in the Atlantic World.* Berkeley: University of California Press, 2009.

Carretta, Vincent. "Equiano's Paradise Lost: The Limits of Allusion in Chapter Five of *The Interesting Narrative.*" In *Imagining Transatlantic Slavery,* edited by Cora Kaplan and John Oldfield, 79–95. Basingstoke: Palgrave Macmillan, 2010.

———. *Equiano, the African: Biography of a Self-Made Man.* Athens: University of Georgia Press, 2005.

Carson, James P. "The Great Chain of Being as an Ecological Idea." In *Animals and Humans: Sensibility and Representation, 1650–1820,* edited by Katherine M. Quinsey, 99–118. Oxford: Voltaire Foundation, 2017.

Chalker, John. *The English Georgic: A Study in the Development of a Form.* London: Routledge and Kegan Paul, 1969.

Chambers, Ephraim. *Cyclopædia: or, An Universal Dictionary of Arts and Sciences.* 2 vols. London, 1728.

Chatterton, Thomas. "Heccar and Gaira. An African Eclogue." *Court and City Magazine,* February 1770, 53.

Chippaux, Jean-Philippe, and Kate Jackson. *Snakes of Central and Western Africa.* Baltimore: Johns Hopkins University Press, 2019.

Churchill, Awnsham, and John Churchill, eds. *A Collection of Voyages and Travels, Some Now First Printed from Original Manuscripts.* 4 vols. London, 1704. [Known as *Churchill's Collection.*]

———. *A Collection of Voyages and Travels, Some Now First Printed from Original Manuscripts.* 6 vols. 2nd ed. London, 1732. [Known as *Churchill's Collection,* second edition.]

Clarke, Samuel. *A True, and Faithful Account of the Four Chiefest Plantations of the English in America. To Wit, of Virginia. New-England. Bermudus. Barbados.* London, 1670.

Clarkson, Thomas. *An Essay on the Slavery and Commerce of the Human Species, Particularly the African, Translated from a Latin Dissertation, Which Was Honoured with the First Prize, in the University of Cambridge, for the Year 1785, with Additions.* London, 1786.

———. *The History of the Rise, Progress, and Accomplishment of the Abolition of the African Slave Trade.* 2 vols. London, 1808.

Clement, John. "Griffith Hughes: S.P.G. missionary to Pennsylvania and Famous 18th Century Naturalist." *Historical Magazine of the Protestant Episcopal Church* 17 (1948): 151–63.

Cobbett, William. *The Parliamentary History of England. From the Norman Conquest in 1066 to the Year 1803.* 36 vols. London, 1806–20.

Coffey, John. " 'I Was an Eye-witness': John Newton, Anthony Benezet, and the Confession of a Liverpool Slave Trader." *Slavery & Abolition* 43 (2022): 181–201.

Coleman, Deirdre. *Henry Smeathman, the Flycatcher: Natural History, Slavery and Empire in the Late Eighteenth Century.* Liverpool: Liverpool University Press, 2018.

Cowper, William. *The Letters and Prose Writings of William Cowper.* 5 vols. Edited by James King and Charles Ryskamp. Oxford: Clarendon, 1979–86.

———. "The Negro's Complaint." In *The Poems of William Cowper,* edited by John D. Baird and Charles Ryskamp, 1:13–14. Oxford: Clarendon, 1980.

Crandall, Joshua. " 'The Great Measur'd by the Less': The Ethnological Turn in Eighteenth-Century Pastoral." *ELH* 81 (2014): 955–82.

Crone, G. R. "Further Notes on Bradock Mead, Alias John Green, an Eighteenth Century Cartographer." *Imago Mundi* 8 (1951): 69–70.

———. "A Note on Bradock Mead, Alias John Green." *Library,* series 5, 6:1 (June 1951): 42–43.

Crosby, Alfred W. *The Columbian Exchange: Biological and Cultural Consequences of 1492* (1972). Westport, CT: Praeger, 2003.

———. *Ecological Imperialism: The Biological Expansion of Europe, 900–1900* (1984). 2nd ed. Cambridge: Cambridge University Press, 2004.

Curran, Stuart. *Poetic Form and British Romanticism.* Oxford: Oxford University Press, 1986.

Dallett, Francis James Jr. "Griffith Hughes Dissected." *Journal of the Barbados Museum and Historical Society* 23:1 (1955): 3–29.

Darwin, Erasmus. *The Botanic Garden; A Poem, in Two Parts. Part I. Containing the Economy of Vegetation. Part II. The Loves of the Plants. With Philosophical Notes.* 2 vols. London, 1791.

Daston, Lorraine, and Peter Galison. *Objectivity.* New York: Zone Books, 2010.

Datiles, M. J., and P. Acevedo-Rodríguez. "*Caesalpinia Pulcherrima* (Peacock Flower)." CABI Compendium (2014), https://doi.org/10.1079/cabicompendium.10728.

Davis, David Brion. *The Problem of Slavery in the Age of Revolution, 1770–1823.* Ithaca, NY: Cornell University Press, 1975.

———. *The Problem of Slavery in Western Culture.* Ithaca, NY: Cornell University Press, 1966.

Davis, Janae, Alex A. Moulton, Levi Van Sant, and Brian Williams. "Anthropocene, Capitalocene, . . . Plantationocene A Manifesto for Ecological Justice in an Age of Global Crises." *Geography Compass* 13:5 (May 2019), https://doi.org/10.1111/gec3.12438.

Day, Matthew. " 'Honour to Our Nation': Nationalism, *The Principal Navigations* and Travel Collections in the Long Eighteenth Century." In *Richard Hakluyt and Travel Writing in Early Modern Europe,* edited by Daniel Carey and Claire Jowitt, 77–86. Farnham, UK: Ashgate for the Hakluyt Society, 2012.

Delbourgo, James. *Collecting the World: The Life and Curiosity of Hans Sloane.* New York: Allen Lane, 2017.

———. "Slavery in the Cabinet of Curiosities: Hans Sloane's Atlantic World." Yumpu (2007). https://www.yumpu.com/en/document/read/4577869/slavery-in-the-cabinet-of-curiosities-hans-british-museum.

Dellarosa, Franca. *Talking Revolution: Edward Rushton's Rebellious Poetics, 1782–1814.* Liverpool: Liverpool University Press, 2014.

Douglass, Frederick. *Narrative of the Life of Frederick Douglass, an American Slave* (1845). Edited by Houston A. Baker Jr. London: Penguin, 1982.

Drake, Thomas E. *Quakers and Slavery in America.* New Haven, CT: Yale University Press, 1950.

Drescher, Seymour. *Econocide: British Slavery in the Era of Abolition.* Pittsburgh: University of Pittsburgh Press, 1977.

———. *Slavery to Freedom: Comparative Studies in the Rise and Fall of Atlantic Slavery.* London: Macmillan, 1999.

Drew, Erin. *The Usufructuary Ethos: Power, Politics, and Environment in the Long Eighteenth Century.* Charlottesville: University of Virginia Press, 2021.

Dubois, Laurent, David K. Garner, and Mary Caton Lingold, eds. *Musical Passage* (2017). http://www.musicalpassage.org.

Dumas, Paula E. *Proslavery Britain: Fighting for Slavery in an Era of Abolition.* Basingstoke: Palgrave Macmillan, 2016.

Edwards, Bryan. *The History, Civil and Commercial, of the British Colonies in the West Indies.* 2 vols. London, 1793.

Egan, Jim. "The 'Long'd-for Aera' of an 'Other Race': Climate, Identity, and James Grainger's *The Sugar-Cane.*" *Early American Literature* 38:2 (2003): 189–212.

Ellis, Markman. " 'Incessant Labour': Georgic Poetry and the Problem of Slavery." In *Discourses of Slavery and Abolition: Britain and Its Colonies, 1760–1838,* edited by Brycchan Carey, Markman Ellis, and Sara Salih, 45–62. Basingstoke: Palgrave Macmillan, 2004.

Equiano, Olaudah. *The Interesting Narrative of the Life of Olaudah Equiano, or Gustavus Vassa, the African* (1789). Edited by Brycchan Carey. Oxford: Oxford University Press, 2018.

Essai sur l'histoire naturelle de St. Domingue. Paris, 1776.

Estwick, Samuel. *Considerations on the Negroe Cause Commonly So Called, Addressed to the Right Honourable Lord Mansfield.* 2nd ed. London, 1773.

Fairer, David. " 'Where fuming trees refresh the thirsty air': The World of Eco-Georgic." *Studies in Eighteenth-Century Culture* 40 (2011): 201–18.

Farrell, Stephen, Melanie Unwin, and James Walvin, eds. "Thomas Clarkson's African Box." In *The British Slave Trade: Abolition, Parliament and People.* Edinburgh: Edinburgh University Press, for the Parliamentary History Yearbook Trust, 2007.

Felsenstein, Frank. *English Trader, Indian Maid: Representing Gender, Race, and Slavery in the New World. An Inkle and Yarico Reader.* Baltimore: Johns Hopkins University Press, 1999.

Ferdinand, Malcom. *Decolonial Ecology: Thinking from the Caribbean World.* Translated by Anthony Paul Smith. Cambridge: Polity, 2021.

Fisher, James. "The Master Should Know More: Book-Farming and the Conflict over Agricultural Knowledge." *Cultural and Social History* 15:3 (2018): 315–31.

Foy, Anna M. "The Convention of Georgic Circumlocution and the Proper Use of Human Dung in Samuel Martin's *Essay upon Plantership.*" *Eighteenth-Century Studies* 49:4 (2016): 475–506.

———. "Grainger and the 'Sordid Master': Plantocratic Alliance in *The Sugar-Cane* and Its Manuscript." *Review of English Studies* 68:286 (2017): 708–33.

Francklyn, Gilbert. *An Answer to the Rev. Mr. Clarkson's Essay on the Slavery and Commerce of the Human Species, Particularly the African; in a Series of Letters, from a Gentleman in Jamaica, to His Friend in London: Wherein Many of the Mistakes and Misrepresentations of Mr. Clarkson Are Pointed Out.* London, 1789.

Fryer, Peter. *Staying Power: The History of Black People in Britain.* London: Pluto, 1984.

Gascoigne, John. *Joseph Banks and the English Enlightenment: Useful Knowledge and Polite Culture.* Cambridge: Cambridge University Press, 1994.

———. *Science in the Service of Empire: Joseph Banks, the British State and the Uses of Science in the Age of Revolution.* Cambridge: Cambridge University Press, 1998.

Gilmore, John. *The Poetics of Empire: A Study of James Grainger's "The Sugar-Cane" (1764).* London: Athlone, 2000.

Gleanings in Africa; Exhibiting a Faithful and Correct View of the Manners and Customs of the Inhabitants of the Cape of Good Hope. London, 1805.

Gragg, Larry. *Englishmen Transplanted: The English Colonization of Barbados, 1627–1660.* Oxford: Oxford University Press, 2004.

Grainger, James. *The Sugar-Cane: A Poem.* London, 1764.

Grant, Hester. *The Good Sharps: The Brothers and Sisters Who Remade Their World.* London: Chatto and Windus, 2020.

Gray, Richard. *Black Christians and White Missionaries.* New Haven, CT: Yale University Press, 1990.

Grove, Richard H. *Green Imperialism: Colonial Expansion, Tropical Island Edens and the Origins of Environmentalism, 1600–1860.* Cambridge: Cambridge University Press, 1995.

Hair, P.E.H., and Robin Law. "The English in West Africa to 1700." In *The Oxford History of the British Empire,* vol. 1: *The Origins of Empire: British Overseas Enterprise to the Close of the Seventeenth Century,* edited by Nicholas Canny, 241–63. Oxford: Oxford University Press, 1998.

Hakluyt, Richard. *The Principall Navigations, Voiages, and Discoveries of the English Nation.* London, 1589.

Hall, Kim F. *Things of Darkness: Economies of Race and Gender in Early Modern England.* Ithaca, NY: Cornell University Press, 1995.

Hansard. https://hansard.parliament.uk/.

Herschthal, Eric. *The Science of Abolition: How Slaveholders Became the Enemies of Progress.* New Haven, CT: Yale University Press, 2021.

Heyrick, Elizabeth. *Immediate, Not Gradual, Abolition; or, An Inquiry into the Shortest, Safest, and Most Effectual Means of Getting Rid of West-Indian Slavery.* London, 1824.

Hobbes, Thomas. *Leviathan* (1651). 3 vols. Edited by Noel Malcolm, Oxford: Clarendon, 2012.

Hochschild, Adam. *Bury the Chains: The British Struggle to Abolish Slavery.* New York: Houghton Mifflin, 2005.

Hollsten, Laura. "Controlling Nature and Transforming Landscapes in the Early Modern Caribbean." *Global Environment* 1 (2008): 176–91.

Hooke, Robert. "A Discourse of Earthquakes in the Leeward Islands." In *The Posthumous Works of Robert Hooke, M.D, S.R.S. Geom. Prof. Gresh. &c. Containing His Cutlerian Lectures and Other Discourses, Read at the Meetings of the Illustrious Royal Society.* London, 1705.

Hosbey, Justin, Hilda Lloréns, and J. T. Roane. "Introduction: Global Black Ecologies." *Environment and Society: Advances in Research* 13 (2022): 1–10.

Hughes, Griffith. *The Natural History of Barbados in Ten Books, by the Reverend Mr Griffith Hughes, Rector of St Lucy's Parish in the Said Island, and F.R.S.* London, 1750.

Iannini, Christopher P. *Fatal Revolutions: Natural History, West Indian Slavery, and the Routes of American Literature.* Chapel Hill: University of North Carolina Press, 2012.

Irlam, Shaun. " 'Wish You Were Here': Exporting England in James Grainger's *The Sugar-Cane.*" *ELH,* 68:2 (Summer 2001): 377–96.

Jackson, Maurice. *Let This Voice Be Heard: Anthony Benezet, Father of Atlantic Abolitionism.* Philadelphia: University of Pennsylvania Press, 2009.

Jamieson, John. *The Sorrows of Slavery, a Poem.* London, 1789.

Jefferys, Thomas. *A General Topography of North America and the West Indies.* London, 1768.

———. *The Natural and Civil History of the French Dominions in North and South America.* 2 vols. London, 1760.

"John Green, Geographer." The Osher Map Library and Smith Center for Cartographic Education, University of Southern Maine (2012). https://oshermaps.org/special-map-exhibits/percy-map/john-green.

Jordan, Winthrop D. *White over Black: American Attitudes toward the Negro, 1550–1812* (1968). 2nd ed. Chapel Hill: University of North Carolina Press, 2012.

Kevan, Keith D. "Mid-eighteenth-Century Entomology and Helminthology in the West Indies: Dr James Grainger." *Journal of the Society for the Bibliography of Natural History* 8:3 (1977): 193–222.

Krawczyk, Scott. "Mediating Abolition: The Collaborative Consciousness of Liverpool's William Roscoe and James Currie." *Journal for Eighteenth-Century Studies* 34 (2011): 209–26.

Kupperman, Karen Ordahl. Introduction to *A True and Exact History of the Island of Barbados*, edited by Karen Ordahl Kupperman, 1–35. Indianapolis: Hackett, 2011.

Labat, Jean Baptiste. *Nouvelle relation de l'Afrique occidentale.* 5 vols. Paris, 1728.

Law, Robin. "Jean Barbot as a Source for the Slave Coast of West Africa." *History in Africa* 9 (1982): 155–73.

Ligon, Richard. *A True and Exact History of the Island of Barbados.* London, 1657.

———. *A True and Exact History of the Island of Barbados.* 2nd ed. London, 1673.

———. *A True and Exact History of the Island of Barbados.* Edited by Karen Ordahl Kupperman. Indianapolis: Hackett, 2011.

Loar, Christopher F. "Georgic Assemblies: James Grainger, John Dyer, and Bruno Latour." *Philological Quarterly* 97:2 (Spring 2018): 241–61.

Locke, John. *Two Treatises of Government* (1689). Edited by Peter Laslett. Cambridge: Cambridge University Press, 1988.

Long, Edward. *The History of Jamaica. Or, General Survey of the Antient and Modern Stage of That Island: With Reflections on Its Situation, Settlements, Inhabitants, Climate, Products, Commerce, Laws, and Government.* 3 vols. London, 1774.

Lovejoy, Arthur O. *The Great Chain of Being: A Study of the History of an Idea* (1936). 2nd ed. New Brunswick, NJ: Transaction, 2009.

Lovejoy, Paul E. "Autobiography and Memory: Gustavus Vassa Alias Olaudah Equiano, the African." *Slavery & Abolition* 27:3 (2006): 317–47.

———. *Transformation in Slavery: A History of Slavery in Africa.* 3rd ed. Cambridge: Cambridge University Press, 2012.

Mackay, David. *In the Wake of Cook: Exploration, Science and Empire, 1780–1801.* Beckenham, UK: Croome Helm, 1985.

Mackenzie, Henry. *Julia de Roubigné, a Tale, in a Series of Letters.* 2 vols. London, 1777.

Mant, Richard. *The Slave, and Other Poetical Pieces; Being an Appendix to Poems by the Rev. Richard Mant.* Oxford: Oxford University Press, 1807.

Martin, Samuel. *An Essay upon Plantership, Humbly Inscrib'd to All the Planters of the British Sugar-Colonies in America.* 2nd ed. Antigua, 1750.

———. *Essay upon Plantership.* 5th ed. London, 1773.

McNeill, J. R. *Mosquito Empires: Ecology and War in the Greater Caribbean, 1640–1914.* Cambridge: Cambridge University Press, 2010.

Meier, Helmut. *Thomas Clarkson: 'Moral Steam Engine' or False Prophet? A Critical Approach to Three of His Antislavery Essays.* Stuttgart: Ibidem, 2012.

Menard, Russell R. *Sweet Negotiations: Sugar, Slavery, and Plantation Agriculture in Early Barbados.* Charlottesville: University of Virginia Press, 2006.

Milton, John. *Paradise Lost.* 2nd ed. (1674). In *The Riverside Milton,* edited by Roy Flannagan. Boston, MA: Houghton Mifflin, 1998.

Mintz, Sidney. *Sweetness and Power: The Place of Sugar in Modern History.* New York: Viking, 1985.

Montesquieu. *De l'Esprit des loix.* 2 vols. Geneva, 1748.

Moody, Jessica. *The Persistence of Memory: Remembering Slavery in Liverpool, "Slaving Capital of the World."* Liverpool: Liverpool University Press, 2020.

Moore, Francis. *Travels into the Inland Parts of Africa: Containing a Description of the Several Nations for the Space of Six Hundred Miles Up the River Gambia.* London, 1738.

Moore, Wendy. *How to Create the Perfect Wife.* London: Weidenfeld and Nicolson, 2013.

More, Hannah. *Slavery, a Poem.* London, 1788.

Morgan, Jennifer L. *Laboring Women: Reproduction and Gender in New World Slavery.* Philadelphia: University of Pennsylvania Press, 2011.

Morgan, Kenneth, ed. *The British Atlantic Slave Trade.* 4 vols. London: Pickering and Chatto, 2003.

Murphy, Kathleen S. *Captivity's Collections: Science, Natural History, and the British Atlantic Slave Trade.* Chapel Hill: University of North Carolina Press, 2023.

Mylander, Jennifer. "Early Modern 'How-To' Books: Impractical Manuals and the Construction of Englishness in the Atlantic World." *Journal for Early Modern Cultural Studies* 9:1 (2009): 123–45.

Nangle, Benjamin Christie. *The Monthly Review, First Series, 1749–1789: Indexes of Contributors and Articles.* Oxford: Clarendon, 1934.

A New General Collection of Voyages and Travels. 4 vols. London, 1745–47. [Known as *Astley's Collection.*]

O'Brien, Karen. "Imperial Georgic, 1660–1779." In *The Country and the City Revisited: England and the Politics of Culture, 1550–1850,* edited by Gerald MacLean, Donna Landry, and Joseph P. Ward, 160–79. Cambridge: Cambridge University Press, 1999.

O'Connor, Kaori. *Pineapple: A Global History.* London: Reaktion Books, 2013.

Ogilvie, Brian W. "Visions of Ancient Natural History." In *Worlds of Natural History,* edited by H. A. Curry, N. Jardine, J. A. Secord, and E. C. Sparry, 17–32. Cambridge: Cambridge University Press, 2018.

Oldfield, John. *Popular Politics and British Anti-Slavery: The Mobilisation of Public Opinion against the Slave Trade, 1787–1807.* New York: Routledge, 1998.

Oldmixon, John. *The British Empire in America, Containing the History of the Discovery, Settlement, Progress and Present State of All the British Colonies, on the Continent and Islands of America.* 2 vols. London, 1708.

Overton, Mark. *Agricultural Revolution in England: The Transformation of the Agrarian Economy: 1500–1850.* Cambridge: Cambridge University Press, 1996.

Parrish, Susan Scott. *American Curiosity: Cultures of Natural History in the Colonial British Atlantic World.* Chapel Hill: University of North Carolina Press, 2006.

———. "Richard Ligon and the Atlantic Science of Commonwealths." *William and Mary Quarterly* 67:2 (April 2010): 209–47.

Pettigrew, William A. *Freedom's Debt: The Royal African Company and the Politics of the Atlantic Slave Trade, 1672–1752.* Chapel Hill: University of North Carolina Press, 2013.

Philips, John. *Cyder, a poem.* London, 1708.

Pierrot, Grégory. *The Black Avenger in Atlantic Culture.* Athens: University of Georgia Press, 2019.

———. "*Droit du Seigneur,* Slavery, and Nation in the Poetry of Edward Rushton." *Studies in Romanticism* 56 (2017): 15–36.

———. "Sable Warriors and Neglected Tars: Edward Rushton's Atlantic Politics." In *Race, Romanticism, and the Atlantic,* edited by Paul Youngquist, 125–44. London: Routledge, 2013.

Pluymers, Keith. *No Wood, No Kingdom: Political Ecology in the English Atlantic.* Philadelphia: University of Pennsylvania Press, 2021.

Pope, Alexander. "An Epistle from Mr. Pope to Dr. Arbuthnot." In *The Twickenham Edition of the Poems of Alexander Pope,* 2nd ed., edited by John Butt, 3:330–31. New Haven, CT: Yale University Press, 1961.

———. *Essay on Man,* In *The Twickenham Edition of the Poems of Alexander Pope,* 2nd ed., edited by John Butt, 3:289–94. New Haven, CT: Yale University Press, 1961.

Pratt, Mary Louise. *Imperial Eyes: Travel Writing and Transculturation.* New York: Routledge, 1992.

Price, Laurence Marsden. *The Inkle and Yarico Album.* Berkeley: University of California Press, 1937.

Priestley, Joseph. *A Sermon on the Subject of the Slave Trade; Delivered to a Society of Protestant Dissenters, at the New Meeting, in Birmingham.* Birmingham, 1788.

Ramsay, James. *An Essay on the Treatment and Conversion of African Slaves in the British Sugar Colonies.* London, 1784.

———. *An Inquiry into the Effects of Putting a Stop to the African Slave Trade and of Granting Liberty to the Slaves in the British Sugar Colonies.* London, 1784.

———. *A Manual for African Slaves. By the Reverend James Ramsay, M.A.* London, 1787.

———. *A Reply to the Personal Invectives and Objections Contained in Two Answers, Published by Certain Anonymous Persons, to "An Essay on the Treatment and Conversion of African Slaves, in the British Colonies."* London, 1785.

Randhawa, Beccie Puneet. "The Inhospitable Muse: Locating Creole Identity in James Grainger's *The Sugar-Cane.*" *Eighteenth Century* 49:1 (2008): 67–85.

Rath, Richard Cullen. "African Music in Seventeenth-Century Jamaica: Cultural Transit and Transition." *William and Mary Quarterly* 50:4 (1993): 700–726.

Raven, Charles E. *John Ray, Naturalist: His Life and Works.* 2nd ed. Cambridge: Cambridge University Press, 1950.

Ray, John. *Historia plantarum.* 3 vols. London, 1686, 1688, 1704.

———. *Methodus plantarum nova.* London, 1682.

———, ed. *The Ornithology of Francis Willughby of Middleton in the County of Warwick Esq, Fellow of the Royal Society.* 3 vols. London, 1678.

———. *The Wisdom of God Manifested in the Works of the Creation. Being the Substance of Some Common Places Delivered in the Chappel of Trinity-College, in Cambridge.* London, 1691.

Rediker, Marcus. *The Slave Ship: A Human History.* London: John Murray, 2007.

Report of the Lords of the Committee of Council Appointed for the Consideration of All Matters relating to Trade and Foreign Plantations; Submitting to His Majesty's Consideration the Evidence and Information They Have Collected in Consequence of His Majesty's Order in Council Dated the 11th of February 1788, concerning the Present State of the Trade to Africa, and Particularly the Trade in Slaves. London, 1789.

Rice, C. Duncan. *The Rise and Fall of Black Slavery.* London: Macmillan, 1975.

Rochefort, Charles de. *Histoire naturelle et morale des iles Antilles de l'Amerique.* Rotterdam, 1658. Translated by John Davies as *The History of the Caribby-Islands, Viz, Barbados, St Christophers, St Vincents, Martinico, Dominico, Barbouthos, Monserrat, Mevis, Antego, &c in All XXVIII.* 2 vols. London, 1666.

Roscoe, William. *The Wrongs of Africa, a Poem.* 2 vols. London, 1787–88.

Rossignol, Marie-Jeanne, and Bertrand Van Ruymbeke, eds. *The Atlantic World of Anthony Benezet (1713–1784): From French Reformation to North American Quaker Antislavery Activism.* Leiden: Brill, 2016.

Rowland, Peter. *The Life and Times of Thomas Day, 1748–1789: English Philanthropist and Author: Virtue Almost Personified.* Studies in British History, vol. 39. Lampeter, UK: Edwin Mellen, 1996.

Rusert, Britt. "Plantation Ecologies: The Experimental Plantation in and against James Grainger's *The Sugar-Cane.*" *Early American Studies,* 13:2 (2015): 341–73.

Rushton, Edward. *The Collected Writings of Edward Rushton, (1756–1814).* Edited by Paul Baines. Liverpool: Liverpool University Press, 2014.

———. *West-Indian Eclogues.* London, 1787.

Sandiford, Keith. *The Cultural Politics of Sugar: Caribbean Slavery and Narratives of Colonialism.* Cambridge: Cambridge University Press, 2000.

———. *Theorizing a Colonial Caribbean-Atlantic Imaginary: Sugar and Obeah.* London: Routledge, 2011.

Sassi, Jonathan. "Africans in the Quaker Image: Anthony Benezet, African Travel Narratives, and Revolutionary-Era Antislavery." *Journal of Early Modern History* 10 (2006): 95–130.

———. "With a Little Help from the Friends: The Quaker and Tactical Contexts of Anthony Benezet's Abolitionist Publishing." *Pennsylvania Magazine of History and Biography* 135:1 (2011): 33–71.

Schaw, Janet. *Journal of a Lady of Quality.* 3rd ed. Edited by Evangeline Walker Andrews. New Haven, CT: Yale University Press, 1939.

Schiebinger, Londa. *Plants and Empire: Colonial Bioprospecting in the Atlantic World.* Cambridge, MA: Harvard University Press, 2004.

———. *Secret Cures of Slaves: People, Plants, and Medicine in the Eighteenth-Century Atlantic World.* Stanford, CA: Stanford University Press, 2017.

Senior, Emily. *The Caribbean and the Medical Imagination, 1764–1834: Slavery, Disease and Colonial Modernity.* Cambridge: Cambridge University Press, 2018.

Sharp, Granville. *A Representation of the Injustice and Dangerous Tendency of Tolerating Slavery or of Admitting the Least Claim of Private Property in the Persons of Men, in England.* London, 1769.

———. A Tract on the Law of Nature, and Principles of Action in Man. London, 1777.

Shelley, Percy Bysshe. "To S. and C." In *The Major Works,* edited by Zachary Leader and Michael O'Neill, 443–44. Oxford: Oxford World's Classics, 2009.

Sheridan, Richard B. "Samuel Martin, Innovating Sugar Planter of Antigua, 1750–1776." *Agricultural History* 34:3 (1960): 126–39.

Shilstone, E. M. "Rev. Griffith Hughes." *Journal of the Barbados Museum and Historical Society* 19:3 (1952): 102–6.

Shuker, Karl P.N. *Extraordinary Animals Worldwide.* London: Robert Hale, 1991.

Shyllon, Folarin. *James Ramsay: The Unknown Abolitionist.* Edinburgh: Canongate, 1977.

Silva, Cristobal. "Georgic Fantasies: James Grainger and the Poetry of Colonial Dislocation." *ELH* 83 (2016): 127–56.

Sinha, Manisha. *The Slave's Cause: A History of Abolition.* New Haven, CT: Yale University Press, 2016.

Slave Voyages Database. https://slavevoyages.org/.

Sloane, Hans. *A Voyage to the Islands Madera, Barbados, Nieves, S. Christophers and Jamaica, with the Natural History of the Herbs and Trees, Four-Footed Beasts, Fishes, Birds, Insects, Reptiles, &c. of the Last of Those Islands; to Which is Prefix'd an Introduction, Wherein Is an Account of the Inhabitants, Air, Waters, Diseases, Trade, &c. of That Place, with Some Relations concerning the Neighbouring Continent, and Islands of America.* 2 vols. London, 1707, 1725.

Smith, Adam. *An Inquiry into the Nature and Causes of the Wealth of Nations* (1776). 2 vols. Edited by R. H. Campbell and A. S. Skinner. Oxford: Oxford University Press, 1976.

Smith, John Samuel. *A letter from Capt. J. S. Smith to the Revd Mr Hill on the State of the Negroe Slaves. To Which Are Added an Introduction, and Remarks on Free Negroes, &c. by the Editor.* Edited by James Ramsay. London, 1786.

Smith, William. *A Letter to William Wilberforce, Esq., MP, on the Proposed Abolition of the Slave Trade, at Present under the Consideration of Parliament.* London, 1807.

Smith, William. *A Natural History of Nevis, and the Rest of the English Leeward Charibee Islands in America.* Cambridge, 1745.

Smith, William. *A New Voyage to Guinea: Describing the Customs, Manners, Soil, Manual Arts, Agriculture, Trade, Employments, Languages, Ranks of Distinction, Climate, Habits, Buildings, Education, Habitations, Diversions, Marriages, and Whatever Else Is Memorable among the Inhabitants.* London, 1744.

Snelgrave, William. *A New Account of Some Parts of Guinea, and the Slave-Trade.* London, 1734.

Snyder, Terri L. *The Power to Die: Slavery and Suicide in British North America.* Chicago: University of Chicago Press, 2015.

Soderlund, Jean. *Quakers and Slavery: A Divided Spirit.* Princeton, NJ: Princeton University Press, 1985.

Some Modern Observations upon Jamaica: As to Its Natural History, Improvement in Trade, Manner of Living, &c. By an English Merchant. In *Whartoniana: or, Miscellanies, in Verse and Prose. By the Wharton Family, and Several Other Persons of Distinction,* edited by Philip Wharton, Duke of Wharton. 2 vols. London, 1727.

Sparrman, Anders. *A Voyage to the Cape of Good Hope, towards the Antarctic Polar Circle, and round the World: But Chiefly into the Country of the Hottentots and Caffres, from the Year 1772, to 1776.* 2 vols. London, 1785.

Stearns, Raymond Phineas. *Science in the British Colonies of America.* Urbana: University of Illinois Press, 1970.

Stedman, John Gabriel. *Narrative of a Five Years Expedition against the Revolted Negroes of Surinam, in Guiana, on the Wild Coast of South America; from the Year 1772, to 1777: Elucidating the History of That Country and Describing Its Productions. Viz. Quadrupedes, Birds, Fishes, Reptiles, Trees, Shrubs, Fruits, &*

Roots; with an Account of the Indians of Guiana & Negroes of Guinea. 2 vols. London, 1796.

Stein, Mark. "Who's Afraid of Cannibals? Some Uses of the Cannibalism Trope in Olaudah Equiano's *Interesting Narrative.*" In *Discourses of Slavery and Abolition: Britain and Its Colonies, 1760–1838,* edited by Brycchan Carey, Markman Ellis, and Sara Salih, 96–107. Basingstoke: Palgrave Macmillan, 2004.

Stevenson, Jane. "Richard Ligon and the Theatre of Empire." In *Shaping the Stuart World 1603–1714: The Atlantic Connection,* edited by Allan Macinnes and Arthur H. Williamson, 285–309. Leiden: Brill, 2006.

Stewart, George. *Reflections, Moral and Political.* 2 vols. Edinburgh, 1787.

Stockdale, Percival. *A Letter from Percival Stockdale to Granville Sharp Esq. Suggested to the Authour, by the Present Insurrection of the Negroes in the Island of St. Domingo.* Durham, n.d. [1791?].

Stott, Anne. *Hannah More: The First Victorian.* Oxford: Oxford University Press, 2003.

Stott, Rebecca. *Darwin's Ghosts: In Search of the First Evolutionists.* London: Bloomsbury, 2012.

Swaminathan, Srividhya. *Debating the Slave Trade: Rhetoric of British National Identity, 1759–1815.* Farnham, UK: Ashgate, 2009.

———. "(Re)Defining Mastery: James Ramsay versus the West Indian Planter." *Rhetorica: A Journal of the History of Rhetoric* 34:3 (2016): 301–23.

Thomas, Hugh. *The Slave Trade: History of the Atlantic Slave Trade, 1440–1870.* New York: Simon and Schuster, 1997.

Thomas, Steven W. "Doctoring Ideology: James Grainger's *The Sugar Cane* and the Bodies of Empire." *Early American Studies* 4:1 (2006): 78–111.

Thomson, Keith. *Before Darwin: Reconciling God and Nature.* New Haven, CT: Yale University Press, 2005.

Tryon, Thomas. *Friendly Advice to the Gentlemen-Planters of the East and West Indies.* London, 1684.

Turnbull, Gordon. *Letters to a Young Planter; or, Observations on the Management of a Sugar-Plantation.* London, 1785.

Uglow, Jenny. *The Lunar Men: The Friends Who Made the Future.* London: Faber and Faber, 2002.

Wadström, Carl Bernhard. *An Essay on Colonization, Particularly Applied to the Western Coast of Africa, with Some Free Thoughts on Cultivation and Commerce; Also Brief Descriptions of the Colonies Already Formed, or Attempted, in Africa, Including Those of Sierra Leona and Bulama.* 2 vols. London, 1794.

———. *Observations on the Slave Trade, and a Description of Some Part of the Coast of Guinea, during a Voyage, Made in 1787 and 1788, in Company with Doctor A. Sparrman and Captain Arrehenius.* London, 1789.

Walvin, James. *Sugar: The World Corrupted, from Slavery to Obesity.* London: Robinson, 2017

———. *The Zong: A Massacre, the Law, and the End of Slavery.* New Haven, CT: Yale University Press, 2011.

Watts, David. *Man's Influence on the Vegetation of Barbados: 1627 to 1800.* Hull, UK: University of Hull Publications, 1966.

———. "Origins of Barbadian Cane Hole Agriculture." *Journal of the Barbados Museum Historical Society* 32 (1968): 143–51.

———. *The West Indies: Patterns of Development, Culture and Environmental Change since 1492.* Cambridge: Cambridge University Press, 1987.

Webster, Jane. "Collecting for the Cabinet of Freedom: The Parliamentary History of Thomas Clarkson's Chest." *Slavery & Abolition* 38:1 (2017): 135–54.

Wheeler, Roxann. *The Complexion of Race: Categories of Difference in Eighteenth Century British Culture.* Philadelphia: University of Pennsylvania Press, 2000.

White, Charles. *Account of the Regular Gradation in Man, and in Different Animals and Vegetables.* London, 1799.

Williams, Eric. *Capitalism and Slavery.* Chapel Hill: University of North Carolina Press, 1945.

Williams, Raymond. *Keywords: A Vocabulary of Culture and Society.* Glasgow: Fontana, 1976.

Wilson, Ellen Gibson. *Thomas Clarkson: A Biography.* York: William Sessions, 1989.

Wise, Steven M. *Though the Heavens May Fall: The Landmark Trial That Led to the End of Human Slavery.* Cambridge, MA: Da Capo, 2005.

Wood, Marcus. *The Poetry of Slavery: An Anglo-American Anthology.* Oxford: Oxford University Press, 2003.

Woodard, Vincent, Justin A. Joyce, and Dwight McBride. *The Delectable Negro: Human Consumption and Homoeroticism within US Slave Culture.* New York: New York University Press, 2014.

Wordsworth, William. *The Prelude.* Edited by Jared Curtis. In *Complete Works: The Cornell Wordsworth,* edited by Stephen Maxfield Parrish, vol. 10. Ithaca, NY: Cornell University Press, 1983.

Zacek, Natalie. *Guide to the Microfilm Edition of the Papers of Samuel Martin, 1694/5–1776, relating to Antigua from the Collections of the British Library: Introduction to the Papers of Samuel Martin.* Microform Academic Publishers, 2010. https://www.research.manchester.ac.uk/portal/files/51121211/MartinPapers.pdf.

———. "Yeomen or 'Grandy-Men'? Class Struggle and Land Distribution in a West Indian Plantation Society." *Manchester Papers in Economic and Social History* 60 (October 2007). http://www.arts.manchester.ac.uk/subjectareas/history/research/manchesterpapers/.

Index